MIRROR IT

Ian Muir Glass and Glazing Federation
Series Consultant Editor: Bob Tattersall

CONTENTS

COLLINS

Introduction

The use of mirrors to increase natural light and create illusions of longer, wider, and more interesting spaces is not new. We have all seen this done, sometimes enjoying the benefits without realizing how it's been achieved.

What is new is the opening up of the world of mirrors to the DIY enthusiast. New methods of production have made the mirror of today more economical and of far better quality than ever before. New products, such as self-adhesive tiles and robust protective backings, give everyone the opportunity to improve their homes. Mirrors are easy to maintain and, installed properly and looked after, they will far outlast paint and wallpaper decorations.

Generally speaking, the DIY enthusiast is most likely to use mirror tiles, mosaic or panels pre-drilled or fixed with clips. Large pieces of mirror cut to size are extremely heavy and difficult to work with, so if at all doubtful, get your supplier to install them.

This book illustrates what can be achieved and then explains, clearly and simply, the many ways of using mirror in your home.

The world of mirrors is a world of illusions and new dimensions. Tremendous imagination can go into improving even the smallest, most awkwardly shaped room.

Far left *This dark attic bathroom is made to feel larger and lighter with the addition of mirror panels behind the bath.*

Above *A mirror panel fixed against the fence gives the trompe l'oeil effect of another garden beyond. The frame and backing panel helps protect the mirror from the elements.*

Left *Mirror-mosaic framed with bamboo brings interesting reflections and light to this dark corner; the orchid adds to the illusion by disguising the bottom edge.*

TOOLS & MATERIALS

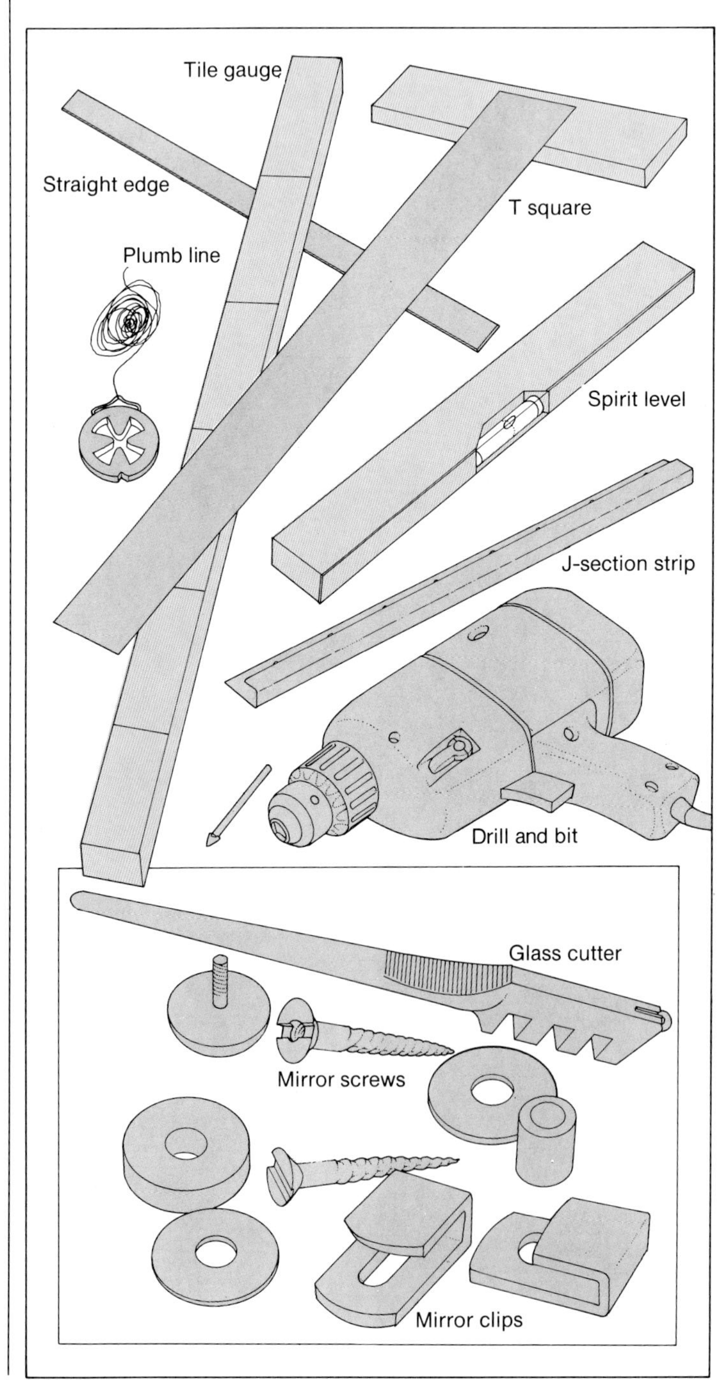

Measuring and Ordering Mirrors

The thickness of mirror used to be labelled by weight per square foot. As the thickness now is stated in millimetres, this is how they compare.

2mm = 16oz/sq ft
3mm = 24oz/sq ft
4mm = 32oz/sq ft
6mm = 48oz/sq ft

Wall mounted mirrors up to one metre square may be 4mm thick. Anything larger should be 6mm.

Always check your measurements carefully, as it is very frustrating to have to return your mirror to be trimmed to size or, worse still, to have had it cut too small.

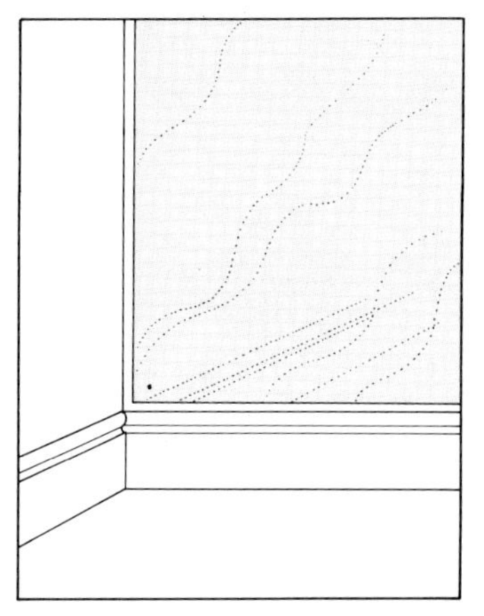

Never run mirrors right up against an adjacent wall or ceiling. Whether using panels or mirror tiles, it is important always to leave a gap so that, if the wall or ceiling isn't straight or if there is movement, the mirror won't be affected. If desired, you can fill the gap around the edges with *mastic* or cover it with *wooden beading*.

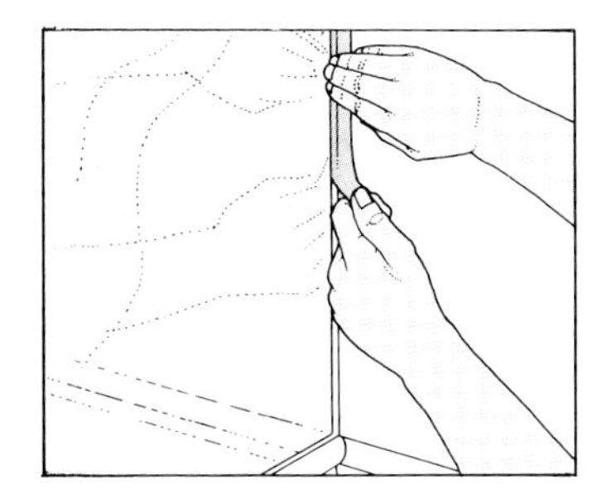

If you have two adjacent walls of mirror, you must leave a space at the junction between the mirror panels. Before fixing the second wall of mirror, run a black glazing tape down the corner. This will black-out the space.

In the case of a large mirror, most suppliers are willing to call at your home to measure. In fact, if they are to install them, they will probably insist. A large mirror is too important a buy to make a mistake with the dimensions.

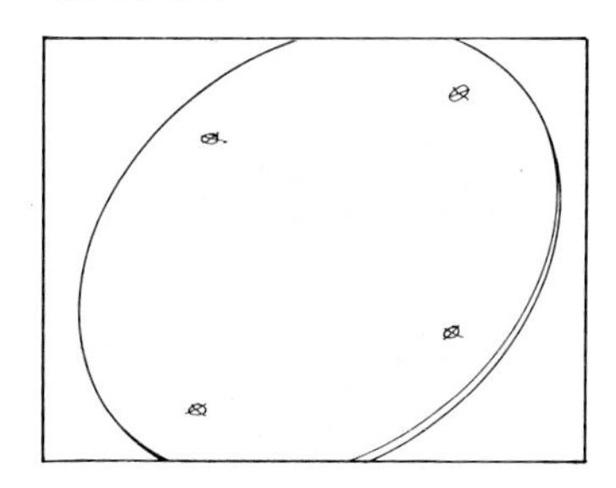

If you wish to mirror an irregular space, make a cardboard template for your glass merchant. Mark on the template any holes that may need to be drilled for light fittings etc.

Tips

Use a mixture of bronze and smoke-tinted mirror tiles for an interesting chequer-board effect.

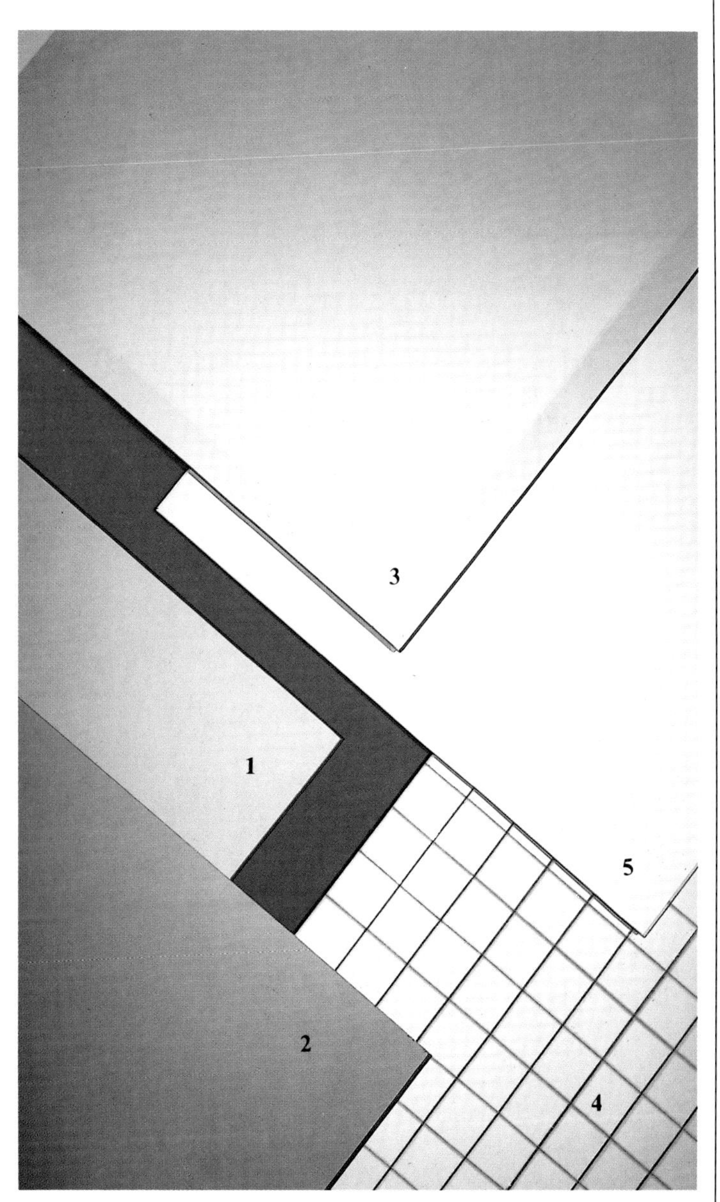

Types of Mirror
1. Bronze-tinted mirror
2. Smoke-tinted mirror
3. Bevelled edge — there are other types of edging, but this is the most common
4. Mosaic sheet — also supplied in tile form
5. Mirror tiles are available in three sizes of plain and tinted mirror, with or without a bevelled edge

Techniques

Additional Protection

When installing mirrors, care must be taken to ensure the mirror back can't be attacked by chemical action from the wall. This is true whether or not you have applied a *polyester covering* to the back of the mirror.

If you have recently moved into a newly built home, or have recently had your walls plastered or rendered, you must not install mirrors until the walls have had an opportunity to dry thoroughly. Alkalis are released from plaster and cement that can damage the mirror's silvering.

Other building materials, such as building blocks, and some adhesives, such as latex cement, contain sulphur that is also harmful to mirrors.

When screw fixing mirrors, it is important that the collars and washers that protect the mirror from contact with the screws are of an inert material. Nylon and polythene can be used, but never use latex or rubber that may contain sulphur compounds.

Mirror tiles that are fixed with adhesive pads must be fixed to a non-porous surface. Surfaces such as plaster, emulsion paint, wood, plywood, chipboard (particle board), and hardboard (fibre board) must first be well sealed with gloss paint – not vinyl. Two coats or more may be necessary and you should allow 72 hours for the paint to dry.

Always use mirror with commercially applied safety backing for sliding mirror wardrobe doors and low-level mirrors.

Although it's possible to cut and drill mirror yourself, it's much easier to take it to your local glazier. Never attempt to cut or drill very large pieces.

Cutting Mirror

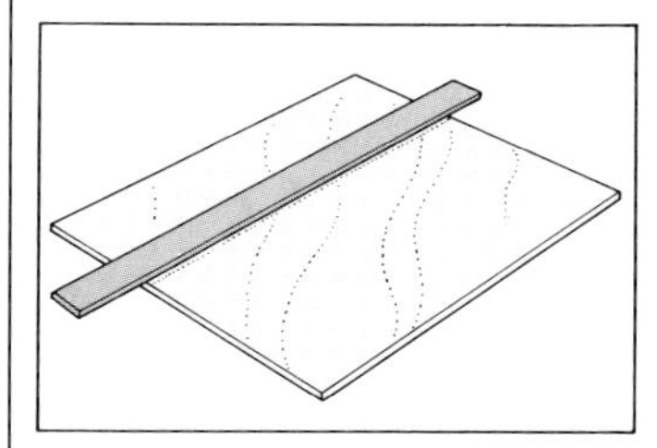

Lay the mirror face up on a clean, flat surface that has been covered with newspapers or a blanket. Measure the cutting line accurately, and position a wooden set square or suitable straight edge at the mark. Don't use a wax pencil for marking as it may affect the efficiency of the glass cutter.

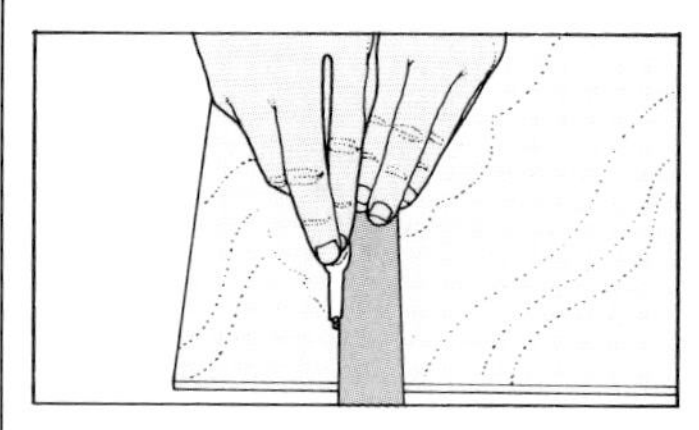

Run the cutter along the straight edge in a single, continuous stroke keeping an even pressure. Don't be tempted to score the glass more than once.

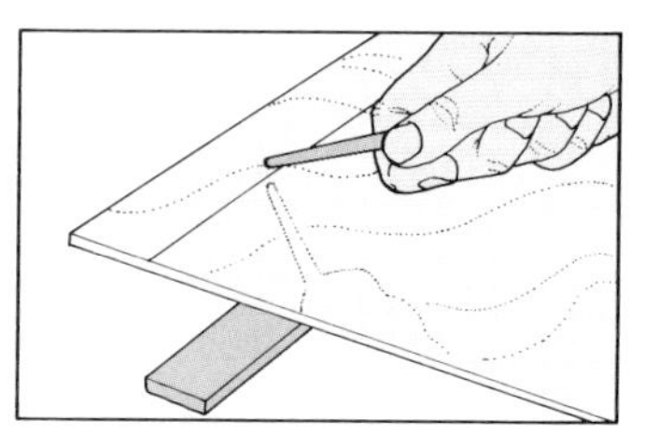

Place the straight edge beneath the mirror and tap firmly with the handle of the glass cutter along the scored line to 'spread the cut'. This stage is important to reduce the risk of an incomplete break along the scored line, which can be difficult to rectify afterwards.

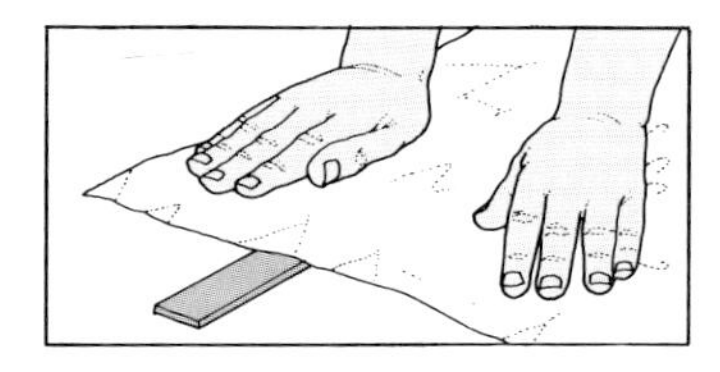

Place the mirror so the cutting line coincides with the straight edge. Place a sheet of newspaper or card on top of the mirror and push downwards firmly to break it along the scored line. Small 'tail pieces' can be nibbled away carefully using pliers or the indentations on the cutter.

Drilling Mirror

Power drills can be used to make holes in mirrors, but they must be capable of running below 500 rpm. If in doubt, use a hand drill. Use a tungsten-tipped spear or spade-point drill bit.

Lay the mirror face down on a smooth, firm surface, such as plywood or chipboard.

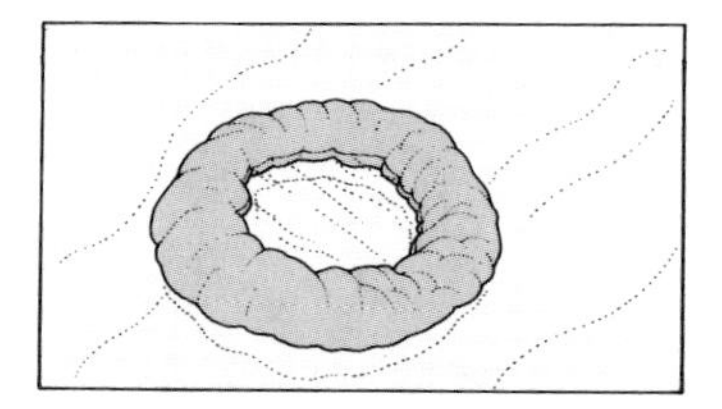

Surround the drilling point with a circle of putty or Plasticine and pour in water as a lubricant.

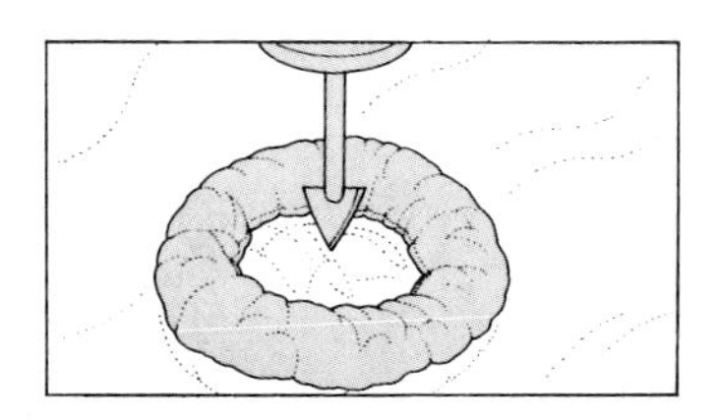

Using low speed and gentle pressure, mark the surface to prevent the drill bit skating off the mark. Continue drilling and removing the bit from the hole often to clear the debris. Don't press too hard or try to hurry it. When the drill is approximately halfway through, turn the mirror over and repeat on the other side.

Smoothing the Edges

You can smooth the edges of a piece of mirror with medium and fine grit abrasive stones of the type used for sharpening knives and woodworking tools. Wear gauntlets to protect your hands and wrists.

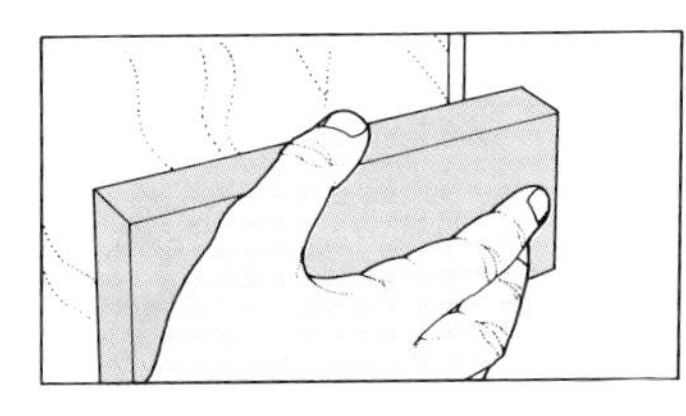

Begin by arrissing the sharp corner between the face and edge of the glass. Use a slightly wetted smooth stone held at 45 degrees in a downward stroke to avoid chipping the edge. Continue until you have a flat diagonal surface at least 1mm wide along all the edges. Repeat the process on the back edges.

Next, use the medium grit stone (again slightly wetted) along the edges. Keep doing this until all shiny patches are gone. Then use the smooth grit stone in the same way.

This process has made the edge of the mirror smooth but not polished.

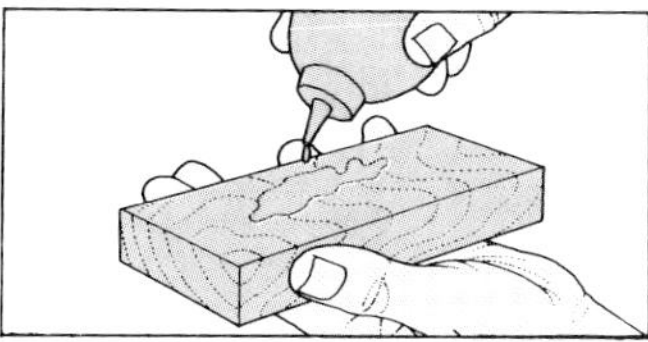

To polish the edge yourself, spread a few drops of household oil on a block of hardwood. Then place the block face down onto a small quantity of jeweller's rouge powder to lightly coat the surface. This is a mild abrasive available at chemists and craft shops.

Use the block along the edges and arrises, occasionally wiping the edge with a cloth to check progress. Although it's possible to achieve a good quality polished edge this way, it does require a lot of time and patience. Glass merchants can smooth and polish edges very quickly and cheaply.

Polishing Scratches

Jeweller's rouge is also useful for polishing out slight scratches in the surface of mirrors. Although glass merchants can do this more quickly, you may have a large mirror installed that would be difficult to move or, perhaps, an old mirror that's seen better days.

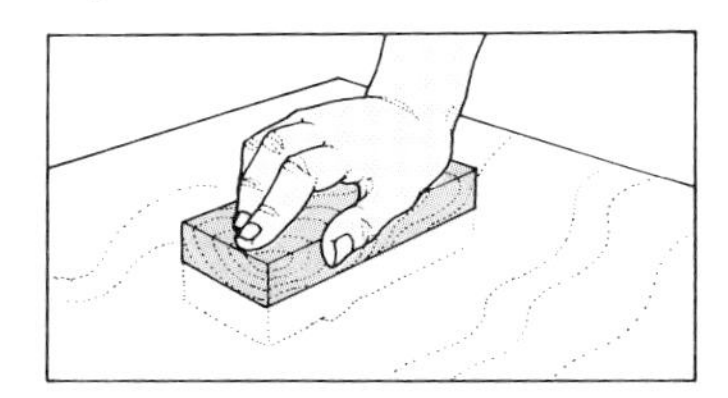

Using a hardwood block as described above, concentrate on the area of the scratch with firm pressure and a circular motion. Alternate with a clean cloth to see the progress and, when you're satisfied, wipe the whole surface of the mirror with the cloth to remove the powder.

Re-silvering

Complete re-silvering of old mirrors is expensive and best left to professional restorers. However, if you have an old mirror that has dark patches beginning to show through, you can brighten these with a simple repair.

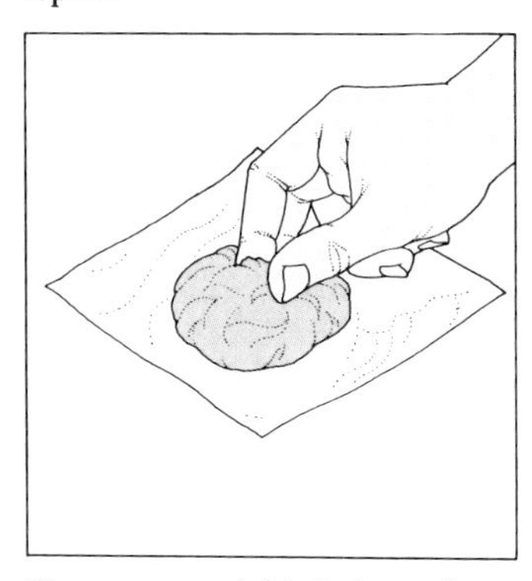

Place an unwrinkled piece of silver paper on a hard surface and rub it vigorously with a clean cloth to polish it.

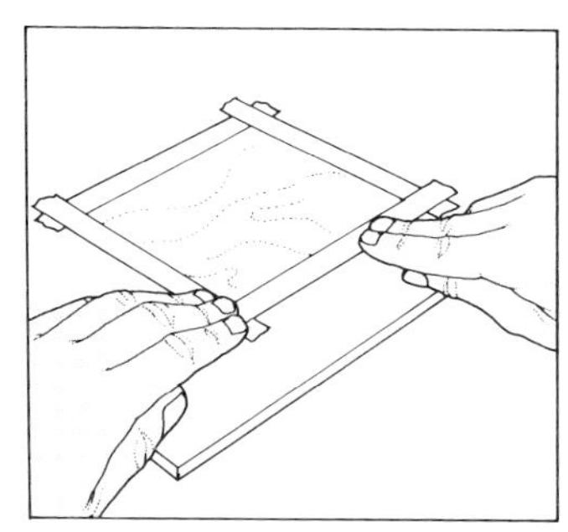

Use adhesive tape to attach the foil over the damaged part of the back of thc mirror.

Protective Backing

Despite protective coatings applied to mirrors during manufacture, the backs are still vulnerable to moisture and chemical attack. It is advisable to apply a protective covering to prevent the mirror's silver degrading.

Originally, lead foil was used to protect the back of mirrors. Nowadays aluminium foil, self-adhesive polyester or polypropylene coverings are available in rolls from glass merchants, and are an investment worth making.

When you use any coverings, the mirror must have a mechanical fixing and not just mastic or adhesive.

Applying the Covering

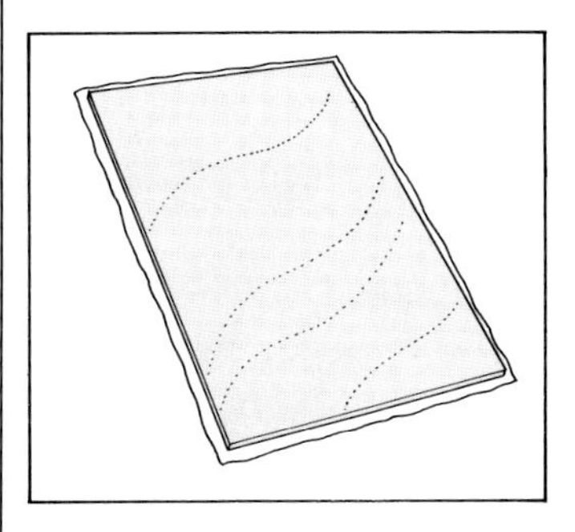

Measure the mirror and buy 5mm extra covering for both length and width. This is to ensure there is enough to cover the whole mirror.

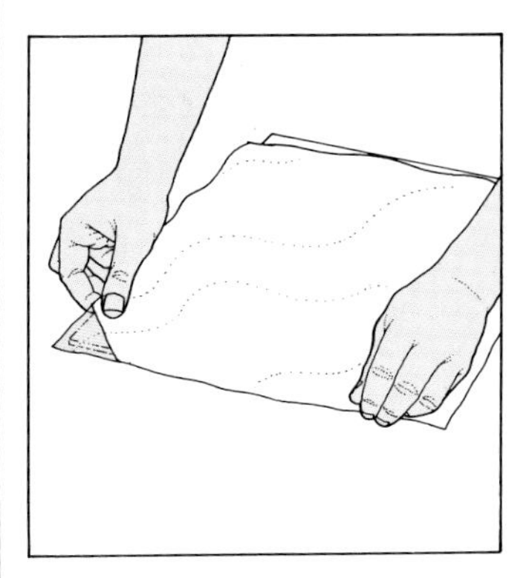

Lay the mirror face down on a soft, flat surface. Peel the protective backing off one corner, and holding it sticky side down, line it up with the edges of the mirror. When it's straight, apply it using a soft, flexible object, such as a plastic ruler.

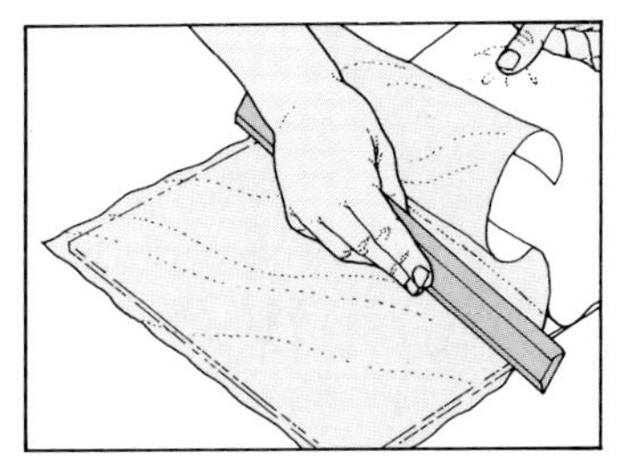

Run the ruler across the covering to apply it to the back of the mirror as you peel off the protective backing.

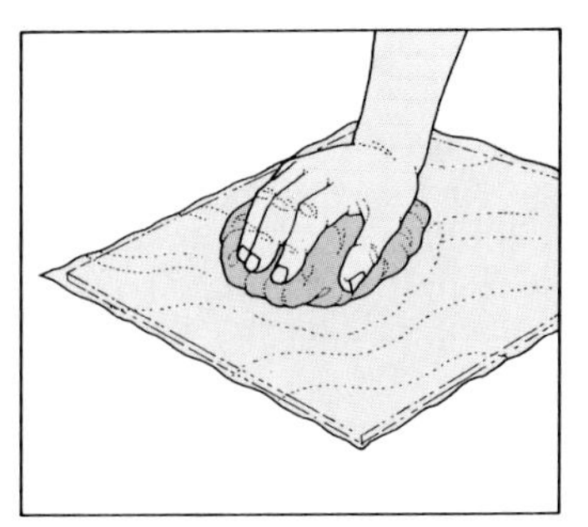

Continue until the whole mirror is covered and then rub gently over the covering with a soft cloth. Don't worry about small air bubbles in the backing, they won't harm the mirror.

If you are covering a mirror larger than the width of the material, simply apply one piece and then overlap another next to it until the entire surface is covered.

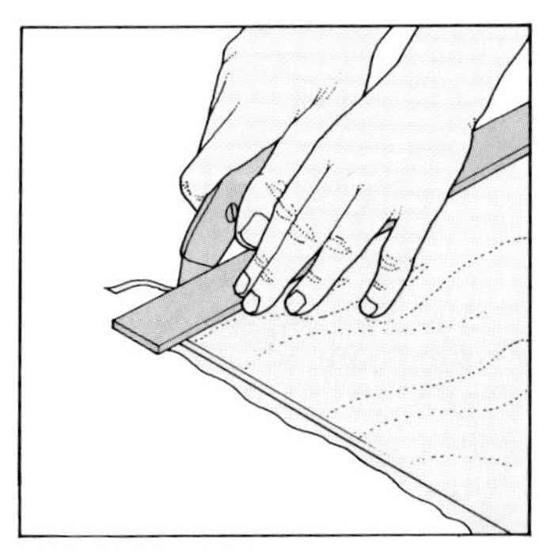

Place a long, straight rule along the edge of the mirror and cut along it with a sharp knife to finish the edge neatly.

If your mirror has pre-drilled holes for screw fixing, use a pencil to carefully break through the covering.

When using mirror in the garden, take care to protect the back and edges from the weather.

MIRROR SAFETY

Once you've decided what mirrors you want and where you want them, it pays to look at the work involved. If the mirrors are very large, or the site high up or awkward to manoeuvre in, try to picture, step by step, how you'll approach it. One tip is to have the edges of a large mirror ground even if it's not strictly necessary for the finished job. It makes handling easier if things get tricky. Finally, if it looks too difficult, you can simply let your local glass merchant do the installation for you. This, at least, makes him responsible for the mirror right up to the end, and what's all in a day's work for him can be worrying for the rest of us.

However, for the jobs you do tackle yourself, here are some things to remember.

Obviously the handling of a heavy and fragile material such as glass requires constant concentration and concern for safety. Don't tackle it when you're tired or in a hurry. This applies whatever size the job; even a small piece of clean cut mirror — say the size of a dinner plate — can, if dropped from head height, cut your leg or foot quite badly. So it makes sense to think through all the risks in each job, and try to minimize them before you start.

Clothing

The most obvious place to start is with *gloves*. These should be of a material tough enough not to be cut through if the glass you're holding should slip, and also give a good grip on the glass. Leather gardening gloves are ideal.

Shoes are another important item. Workboots with toecaps are best, but any stout leather shoes will do. Lightweight canvas and leather casual shoes should be avoided.

As for *clothes*, long sleeves with buttoned cuffs are better than short sleeves (for obvious reasons) and it's a good idea not to wear anything too flowing that could catch on a sharp corner.

Storage

Particularly if you are dealing with large mirrors, you must be sure that you have a safe storage place prepared before taking delivery of the mirror.

Small pieces of mirror that are being kept for use at a later stage should always be stored well away from children, preferably wrapped in corrugated cardboard and labelled clearly. The corners and edges are particularly vulnerable. Apart from the safety angle, it's maddening to measure and order your piece of mirror, only to find it damaged when you are ready to use it.

Some installation methods require holding the mirror against the wall to mark screw holes, and then taking it down again. Be sure to place the mirror safely during the installation process — especially if you have children in the house.

Drilling and Cutting

It's wise not to attempt to drill or cut mirrors yourself. Not only is it very easy to ruin the mirror using amateur tools, but it can also be dangerous. Your glass merchant will be happy to drill and cut your mirrors precisely as you wish.

Ladders

Carrying mirrors of any size up ladders requires experience and training. Do not attempt this by yourself.

Helpers

Never install a mirror without enough help. This should involve enough people to hold the mirror in position, while you mark holes or stand back to see whether it looks straight.

Large mirrors

All mirrors larger than 1 square metre (however they are fixed) must have additional support that is adequate to hold their weight.

Sliding mirror doors

Sliding mirror doors are a potential hazard as there is a chance that someone could fall against the large areas of glass. In view of this, all mirrors for wardrobes must have a commercially applied plastic film that holds the broken fragments of glass together, reducing the risk of serious injury.

Low-level mirrors

The same applies to these as to sliding mirror wardrobe doors.

Be careful not to site mirrors in such a way that they could be mistaken for doorways.

WHERE & HOW TO USE MIRRORS

Take a look at your own home and decide where it might be interesting to 'alter' the dimensions. Perhaps your living room is long and narrow and needs a little more depth, or you may have a gloomy hallway that needs brightening. This part of the book will help you to realize the scope and potential for mirrors.

The first thing to remember is that you are using the reflection in the mirror to create the illusion of additional width, length or height. The reflection also brings light deeper into the room, giving the impression of a greater natural light source.

It is important to grasp that we are speaking of using mirror not as a large ornament, but as a means of creating an illusion. Therefore, the panel of mirror glass should be 'invisible'. It is the reflection of additional wall, floor or ceiling space we want to see, not the mirror itself.

Because it is the reflection that is important, the mirror should have something attractive to reflect. This means the walls, floors and ceilings need special attention — there is no point in reflecting a blank wall. A mirror doubles the impact of an object, so make sure the impact is worthwhile. Try positioning a photograph, print or painting opposite a mirror.

Where space is confined and lighting poor, in alcoves, cupboards and storage areas for example, the addition of a mirror will optimize available light.

Ceilings can be 'lifted' by panelling a part of them using mirrors in a deep frame. The impression of depth can also be achieved by the use of mirrors inset behind a chandelier or distinctive light fitting.

Often too small and badly proportioned, bathrooms are another natural place for mirrors. The feeling of additional space and light creates a stylish atmosphere where one can prepare for the day. Effective results can be achieved using mirror tiles behind the bath.

As far as possible, the mirror should be placed where the occupants of the room will not be constantly seeing their own reflection. The clever use of plants and furnishings will help.

Top *Covering a wall with mirror doubles the apparent length of any room. Because this apparent increase in length comes from reflections, these should be emphasized by the use of ceiling lines, venetian blinds and any wall pattern that will help disguise the edges of the mirror.*

Above *The apparent width of this room is doubled by having an entire wall covered with mirror. Placed at right angles to the window, it has the effect of doubling the window area and dramatically increasing the amount of light in the room. Here the effect is produced by full-length wardrobe doors with the panels carefully concealed.*

Left *Here a relatively cramped study is given a spacious, comfortable feeling using a large mirror. Fitting tightly up against the ceiling, with its base concealed by the bookcase, the mirror's presence is disguised.*

Below *This is an illustration of how, with careful planning, a mirror's presence can be concealed. Room occupants will feel they are in a wider, larger room without being constantly aware of the mirror panels. Flanked by bookshelves, the edge of the mirror cannot be seen, so there is no break in the illusion. Careful positioning of plants prevents room occupants continually seeing their own reflection.*

Above *Mirror has been neatly fitted into this alcove, reflecting shelves and ornaments. Carefully lit, the effect is stunning.*

Left *The side of the bath has been mirrored, along with the end walls, transforming a very small space into a multitude of reflections and angles.*

Top left *This large plain mirror over a fireplace creates a spacious atmosphere by adding depth to the room. This is a traditional position for a large mirror.*

Top right *Fitted into window reveals, mirrors bounce increased daylight into a room. They also give the illusion of a larger window.*

Right *Many halls are long, narrow and badly lit. Here you can see how this can be remedied. The mirror doubles the apparent space, reflects light and produces a welcoming entrance from what was a drab hallway.*

Creating Light & Space

The following instructions show the basic rules that apply when trying to improve the dimensions of a room or, perhaps, accentuating a particular feature. Obviously it's rarely possible to stick rigidly to any set of rules — they must be interpreted for each individual set of circumstances. Having seen and understood these rules, you should be able to judge how best to achieve the result you're after. For example, it is not usually a good idea to place mirrors on opposite walls as it has the disorienting effect of reflecting images eternally. However, bearing this warning in mind, some interior designers have used this effect to produce stunning results.

Width

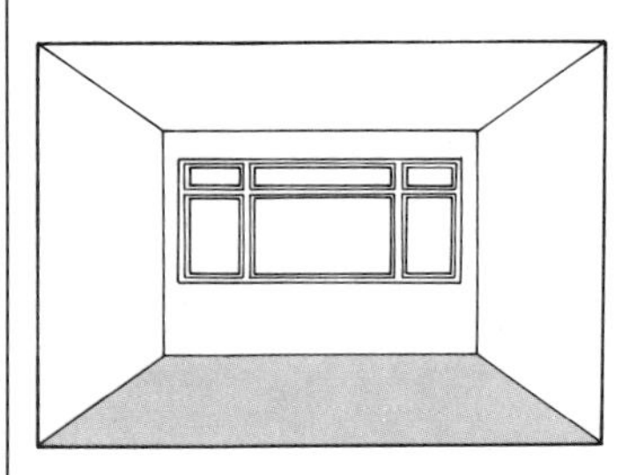

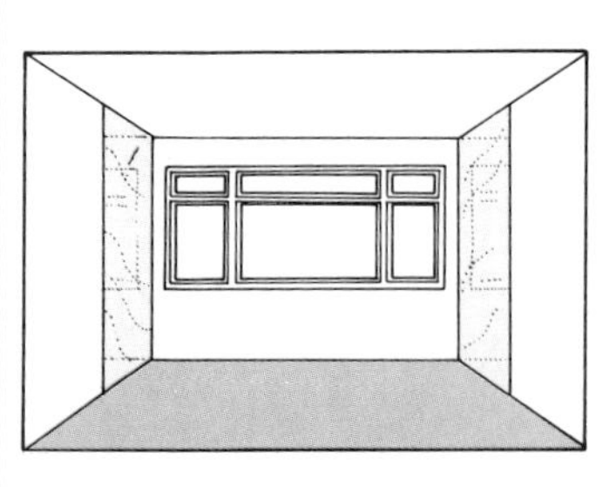

One way of making a room look wider is to fit floor to ceiling mirror panels adjacent and at right angles to the window. This apparent increase in width should be emphasized by the use of venetian blinds and any wall pattern that will help disguise the edge of the mirror.

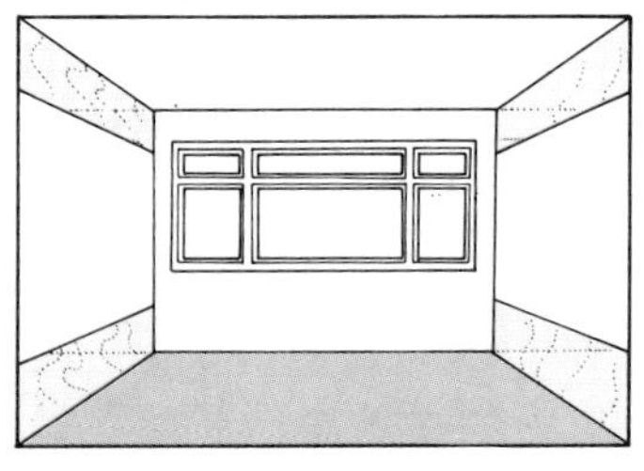

Rooms can also be made to appear wider by fitting long horizontal bands of mirror across the top and bottom of the side walls and giving the centre section a decorative wall treatment.

Fix mirror or mirror tiles at the back of your workbench or kitchen counter. Not only will this give you additional reflected light, but you will be able to see what is happening at the back of an awkward job.

Height

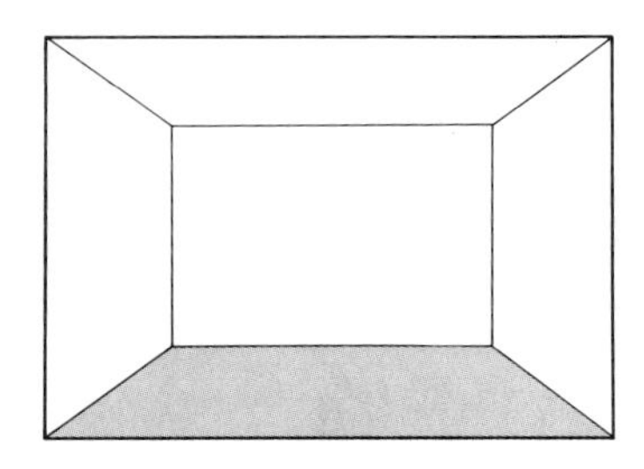

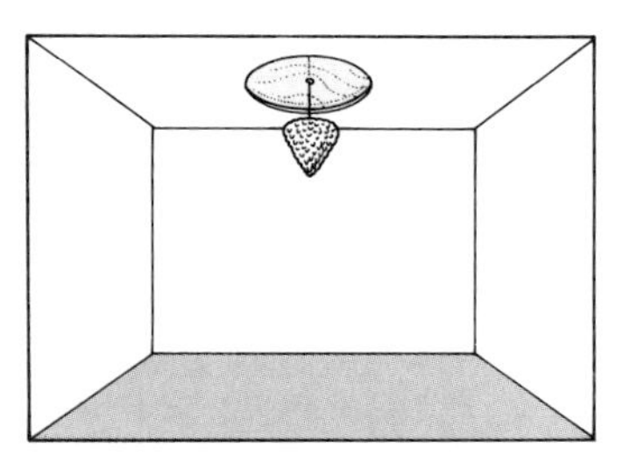

A ceiling can be 'lifted' by panelling a section with mirror within a deep frame. The impression of depth can be strengthened by distinctive light fittings.

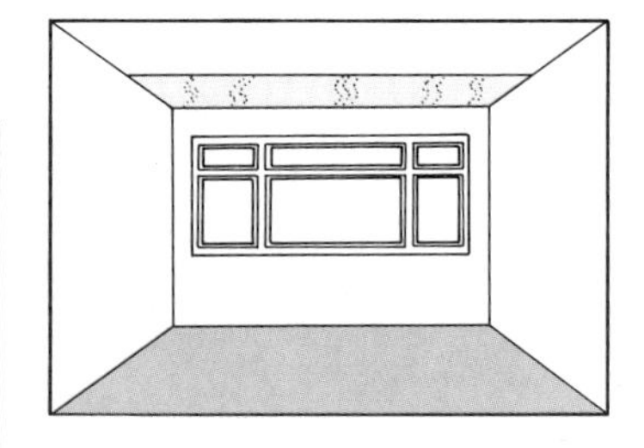

Here is a use of mirrors that gives the impression of greater height and taller windows. A mirror panel is fixed to the ceiling above the window.

Length

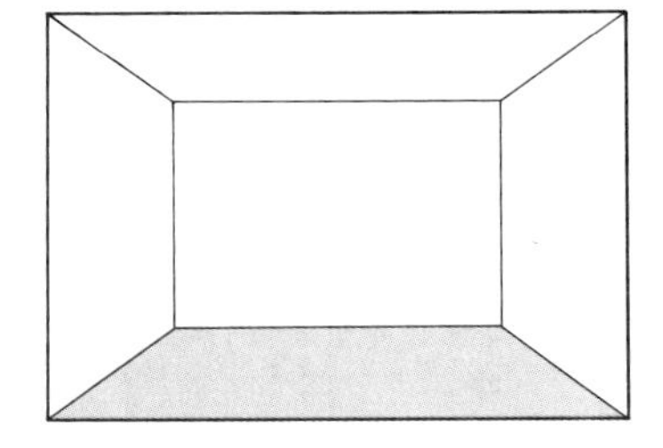

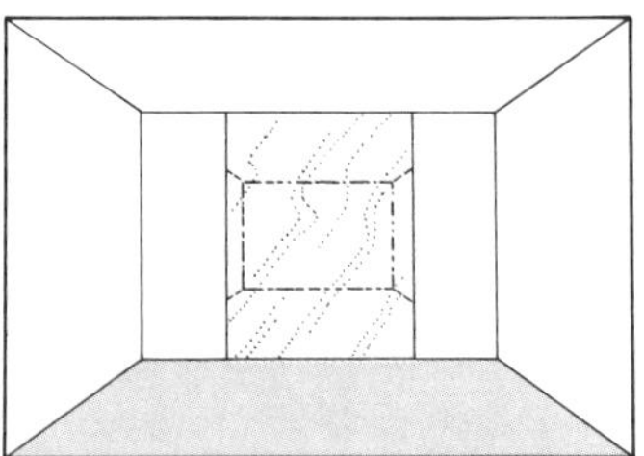

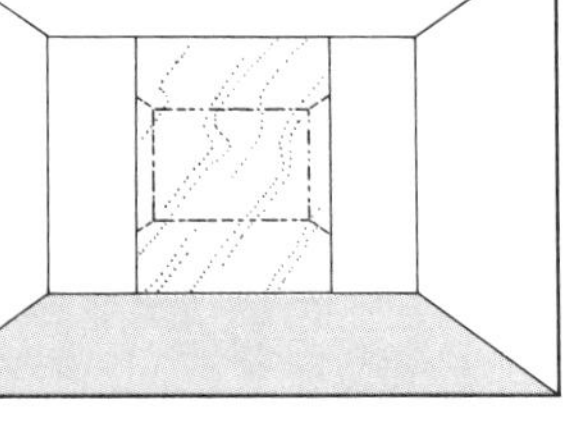

The impression of a much longer room can be obtained by fixing a floor to ceiling mirror panel in the centre of the end wall.

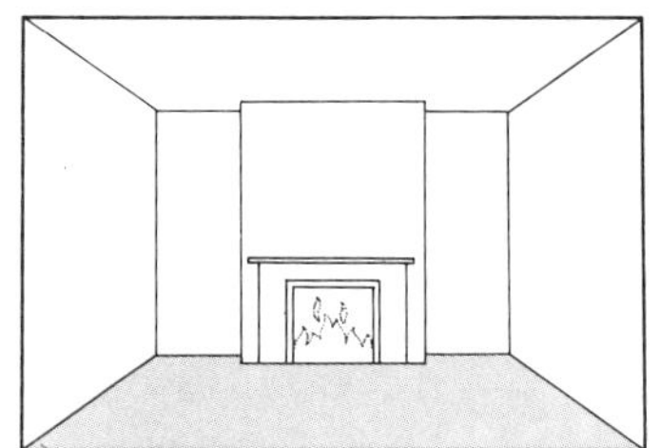

Another way to obtain extra length is to fit two mirror panels in the alcoves either side of a chimney breast.

Light

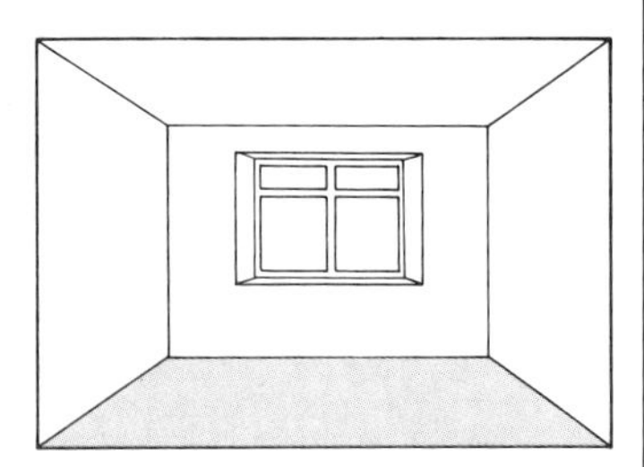

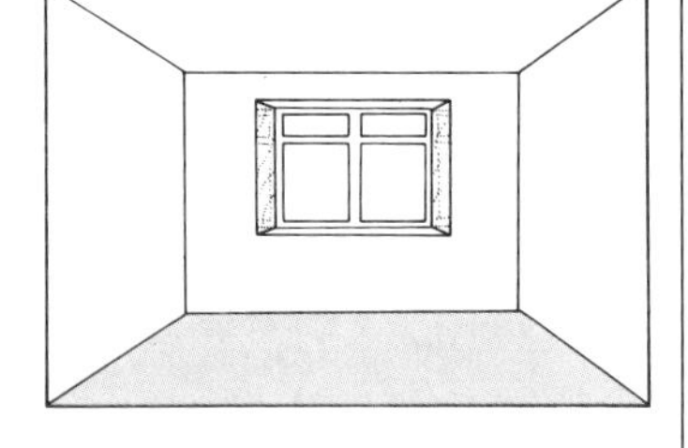

Increased light can be obtained by lining the inner reveals of the windows.

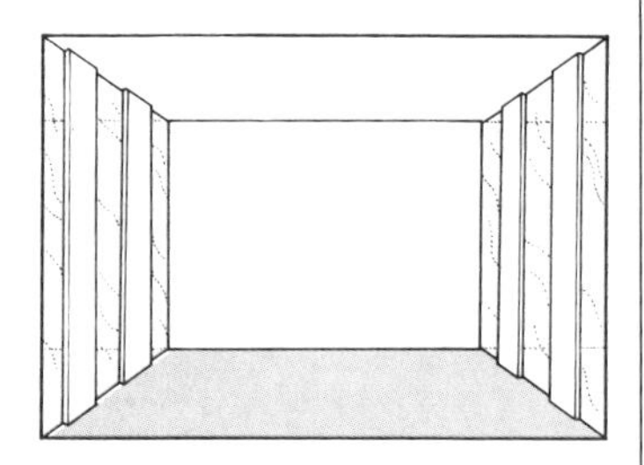

Floor to ceiling panels, spaced at intervals along the walls adjacent to the window, will help light to penetrate into the depths of the room.

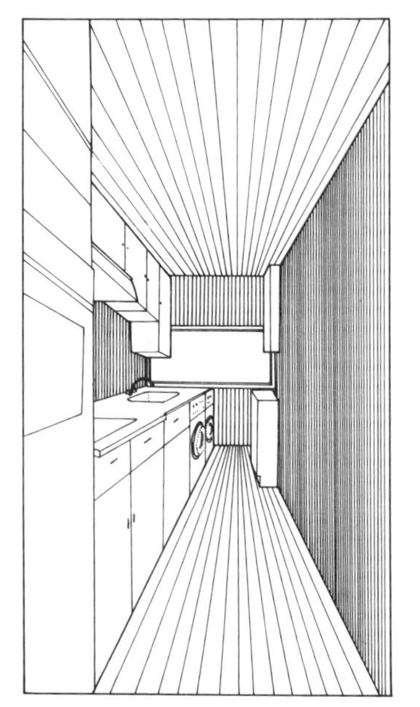

Left *Although quite narrow, the addition of mirror from floor to ceiling on one wall gives the illusion of more width. The diagonally-laid floor tiles also help.*

FIXING MIRRORS

Whichever method you use to mount your mirrors, if the wall is uneven, you will need to produce a flat surface. Otherwise the mirror's reflections will be distorted and larger mirrors can be broken.

If you want to surround a large fixed mirror with a frame, mounting the mirror on blockboard will give you a wider choice of suitable frames and also something to attach the frame to.

Producing a Flat Surface

An even surface can be made by fixing blockboard to the wall by means of 50mm × 25mm timber battens.

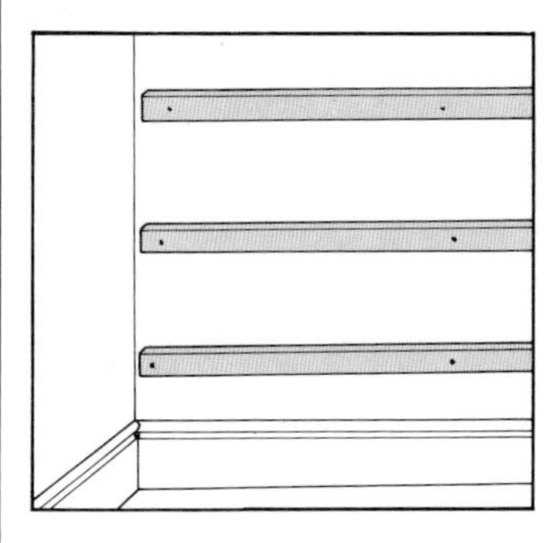

The battens are fixed horizontally to the wall, parallel with the skirting. They should be placed approximately 300mm apart.

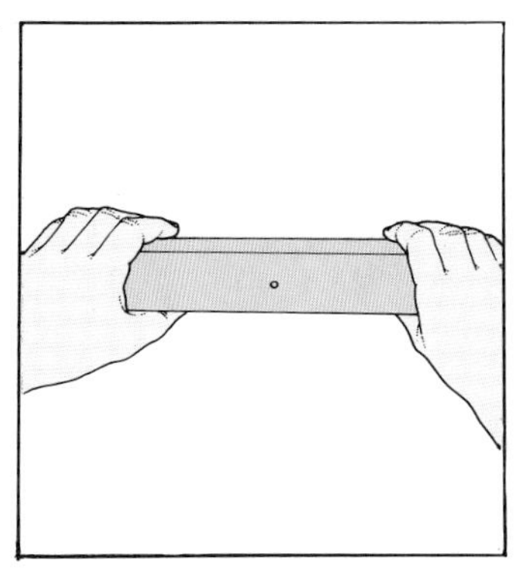

Drill the battens at each end and in the middle, and place them against the wall.

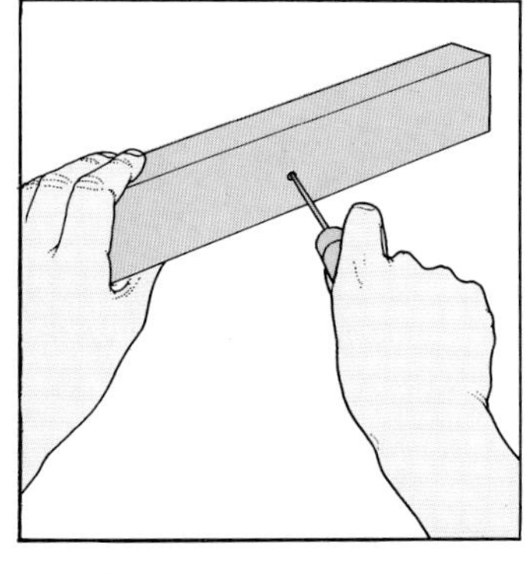

Use a bradawl through the holes to mark their positions on the wall. Mark all the battens at once, numbering them so you can put them up in the right order.

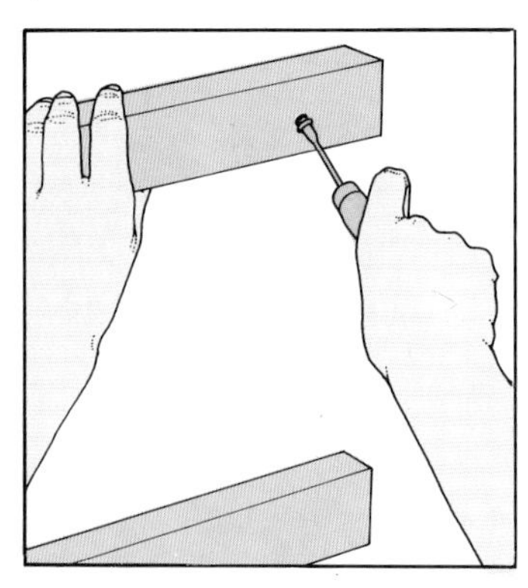

Drill holes in the wall, insert wall plugs, then, picking up the battens in turn, screw them to the wall. Don't screw them too tight initially, as you will need to adjust them.

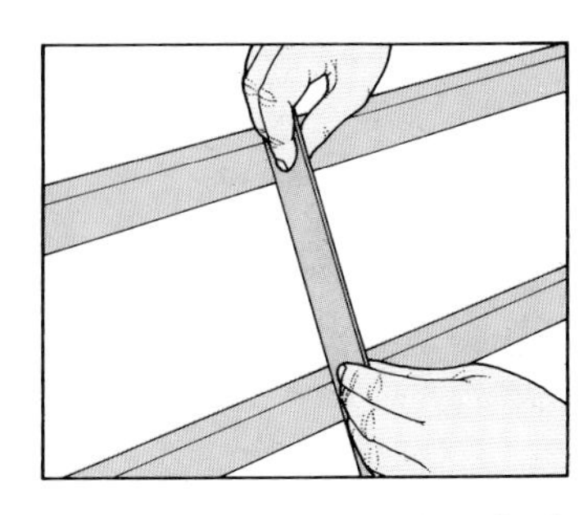

Now, using a straight edge, check the battens are flat.

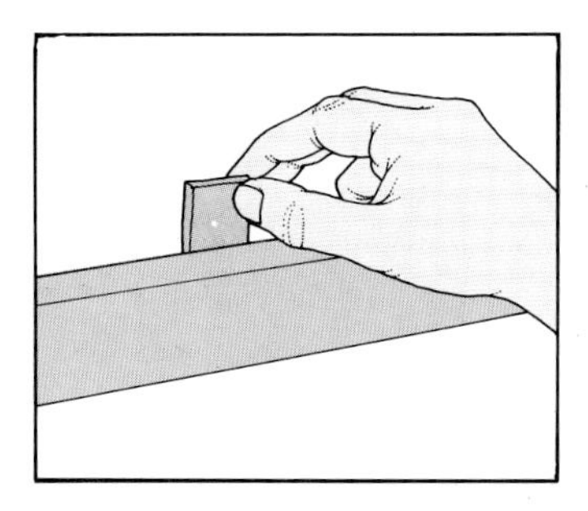

Where you find they are too low, use small pieces of wood as shims to lift them up. High points can be planed down. When you're confident that the structure is as flat as possible, tighten up the fixing screws.

Now you can cut the blockboard to size and screw it to the battens. Then seal the surface with *gloss paint* — not vinyl. Apply two or more coats, if necessary, and allow at least 72 hours to dry.

Clip Fixing

The simplest method of attaching mirrors to walls is to use special clips (see Tools and Materials). With this method the mirror rests on two fixed clips, and is secured by sliding clips on the other three sides. This is a suitable method of fixing small mirrors, such as those over basins in bathrooms.

As all the mirror's edges are exposed, it is essential that they are properly finished. The glass merchant should be asked to arris, grind or polish the edges.

Remember to check that your wall is flat, and make sure it's sound and strong enough to take the weight of the mirror.

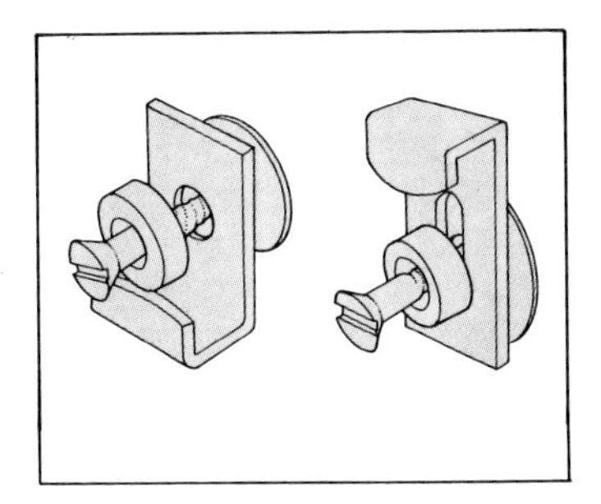

There will be two fixed clips for use on the bottom edge of the mirror, and four sliding clips, to be used around the other three sides.

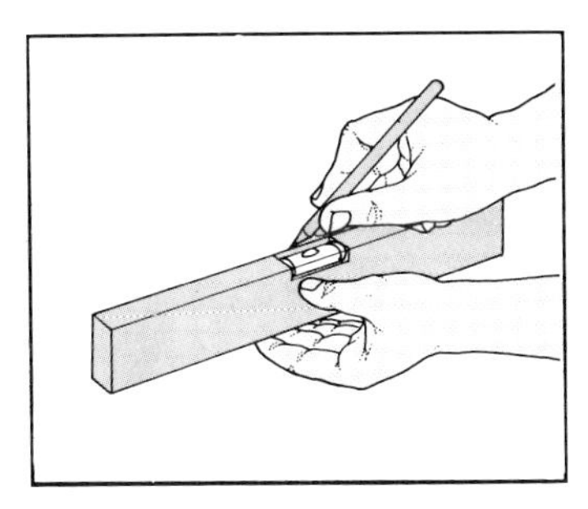

Decide where to fix the mirror and, using a spirit level, draw a horizontal line on the wall where the bottom edge will be. Also, mark the position of the bottom corners of the mirror on the line.

Mark two fixing holes 13mm above the bottom line and approximately one-quarter of the way in from each corner. Drill and plug the holes.

Fix the clips with screws so the thin metal washer is against the wall, then the clip, and then the thick nylon washer. The screw head is totally countersunk into this large washer. Be careful not to overtighten the screws, they should be firm — no tighter.

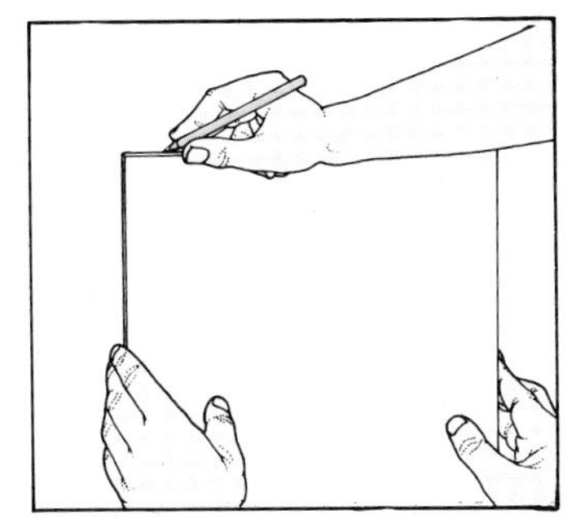

While a helper rests the mirror on the bottom clips, pencil around the other three sides. Remove the mirror and place it safely out of the way.

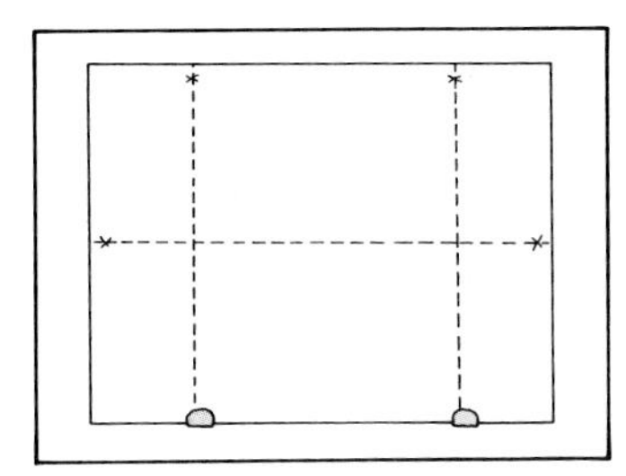

Mark the holes for the three sliding clips. The two clips on the top should be spaced as the bottom ones — one quarter in from each side but not less than 75mm from the corner. Mark the holes 19mm below the line of the top edge of the mirror. The clip at each side should be halfway up the side of the mirror and, again, mark the hole 19mm in from the line.

Drill and plug the four remaining fixing holes and attach the sliding clips and washers. Tighten the screws so the clips are firmly held in place, but are still easy to slide.

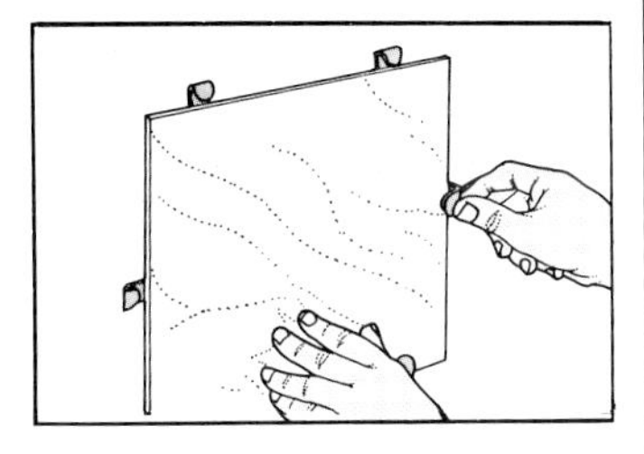

Now, with the top and side clips slid outwards from the centre, place the mirror on the bottom clips and push the sliding clips into position so they retain the mirror without undue stress.

If you find that any of the clips don't fit easily over the mirror, the wall is probably uneven. To compensate for this, use washers behind (against the wall) the clips until all fit properly.

Screw Fixing

Screw fixing is an efficient and effective means of attaching small and medium size mirrors. The special screws with attractive coverheads can be made a feature of the installation. Use this method or clip fixing where damp conditions prevail — e.g. kitchens or bathrooms.

Your local glass merchant will drill the mirror for you and you can specify precisely where the holes are required. Bear in mind that the holes should not be less than 50mm from the edge of the mirror, and that the holes must be large enough to accommodate the screws and the plastic collars that protect the inside of the holes.

Preparation

As with clip fixing, make sure the wall is sound and strong enough to take the weight of the mirror. Also the surface should be sealed with gloss paint.

If you're fixing to a door or something other than a wall, make sure the screws aren't so long they will come out the other side.

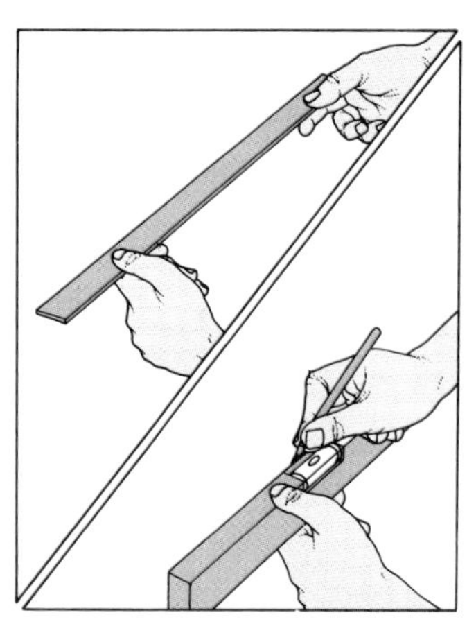

Use a straight edge to check the flatness of the wall. Slight uneveness can be compensated for by using extra spacers, but if the wall is very out of true, use the blockboard method (page 16) to produce a flat surface.

Decide where the mirror is to be installed and mark the baseline using a spirit level.

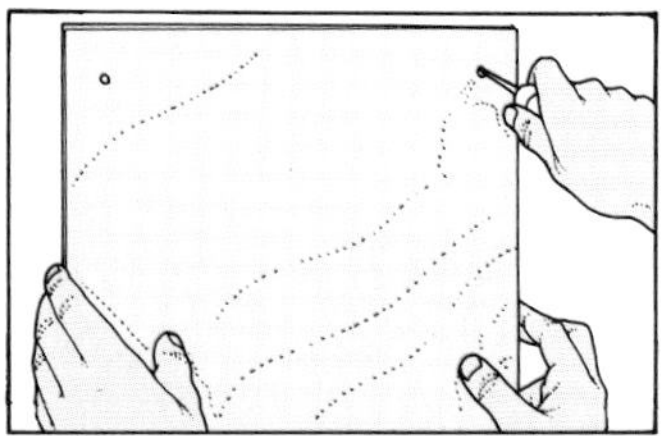

While a helper holds the mirror firmly against the wall, mark the screw holes with a bradawl. Be sure to mark carefully in the centre of the holes. Then put the mirror safely to one side.

Drill the holes exactly on the marks and it is important to drill straight into the wall — not at an angle.

Fit plugs into the holes and hold up the mirror to check that the holes match exactly.

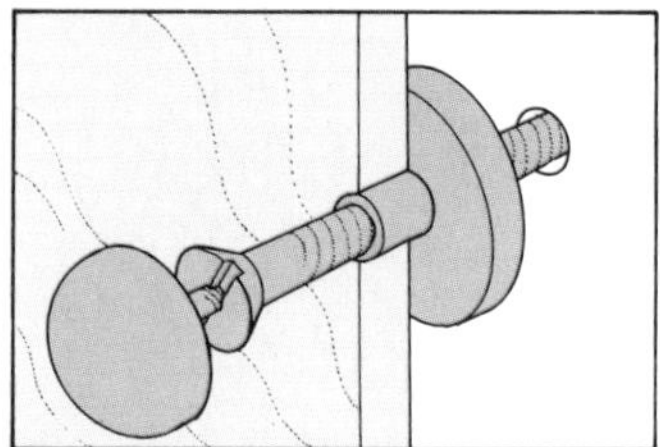

Begin to fix the screws but remember it is essential there is no glass-to-metal contact. Use the recommended washers on both sides of the mirror, and a collar between the screw shank and the inside of the mirror holes.

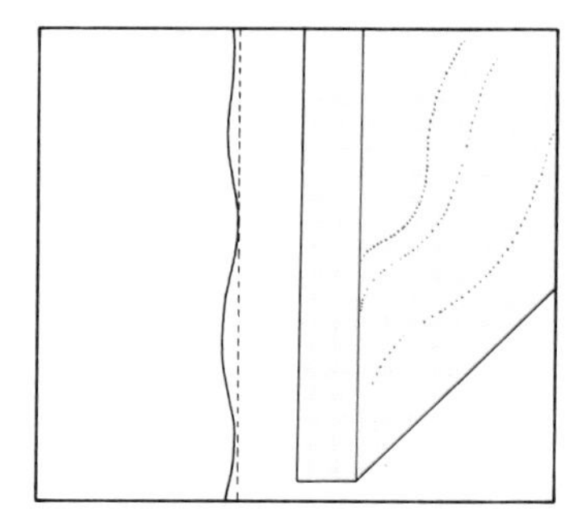

It is important that there is always an airspace of *at least 3mm* between the mirror and wall to allow the mirror to breathe. If the wall is uneven, the gap should be increased to allow for the high spots.

Now carefully tighten the screws in turn. To tighten a screw that is going into the wall crooked would crack the mirror so make sure they are going in straight.

Continue turning the screws until their heads reach the washer. Don't overtighten them. This would distort the reflection and could break the mirror.

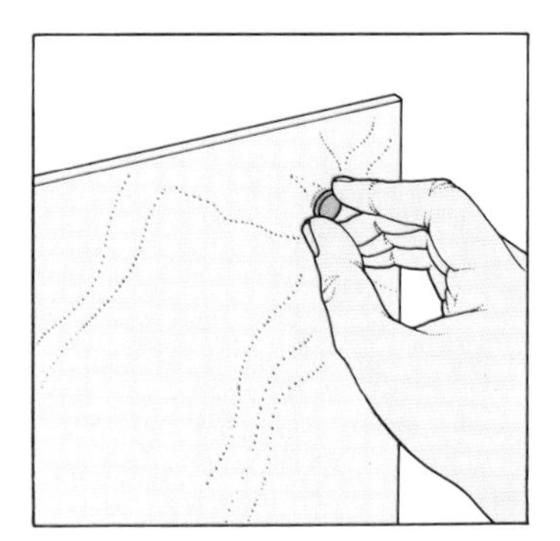

When all the screws are in place, screw on the cover-heads.

If, at any time, you realize that any of the holes are incorrect, you should stop immediately. You can fill the holes, drill new ones and start again.

Fixing With Compound & Silicone

The most subtle fixing method for mirrors is by using fixing compound together with silicone. With no clips or screws to betray it, the mirror's presence can be concealed more effectively. It is particularly important to follow all the fixing instructions carefully to ensure a sound, secure installation. There must always be mechanical support in addition to the compound and silicone, whether provided by a skirting board, floor, worktop or J-section strip.

Preparation

Careful preparation is essential. Begin by checking the mounting surface. Make sure it is sound, not at all damp, and that there is no peeling plaster. If you are in any doubt at all, don't use this method to mount the mirror.

It is especially important with this method of fixing to seal the surface. Whether you're mounting on wood or plaster, your glass merchant will be able to recommend a suitable glazing sealer. Follow the manufacturer's instructions carefully — especially regarding drying times.

Fixing

Once you are sure the wall is well sealed and sound, you must ensure you have a sufficient mounting platform for the mirror. This is to take the weight of the mirror; the compound and silicone are only to hold it to the wall.

If your mirror is floor to ceiling, you may find it easiest to rest it on the skirting board. If you rest it on the floor, you'll need to remove part of the skirting board.

Tips

Never use any adhesives except those recommended for use on mirrors, or the silvering may discolour.

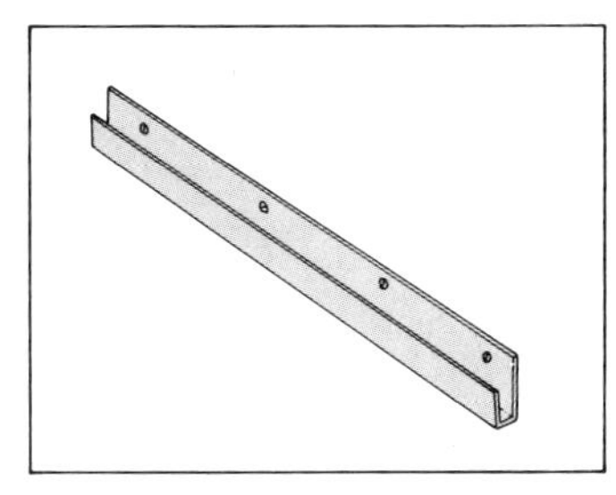

If the mirror isn't to rest on the skirting, worktop or other substantial ledge, then the easiest way to provide support is to use *J-section aluminium strip.*

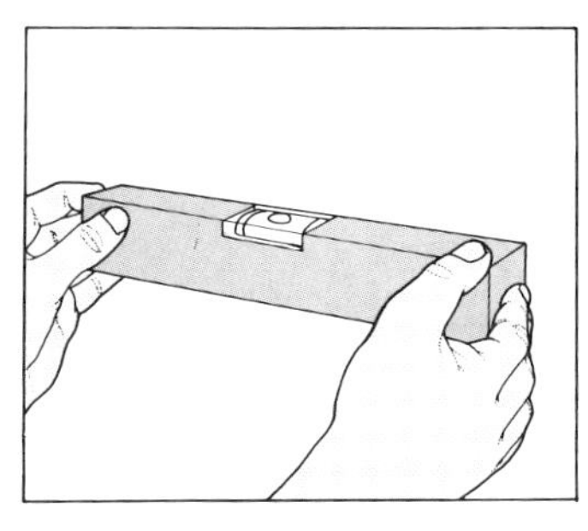

Using a spirit level, position the strip and mark sufficient screw holes for the weight of the mirror. Then drill and plug the holes and fix the strip using at least 30mm screws.

Once all has been prepared, you can arrange delivery of the mirror. This is the time to enlist the help of a friend — it is necessary to have an assistant to help fix the mirror. Think carefully about its size and don't take any chances. Sufficient manpower to raise the mirror safely is absolutely essential.

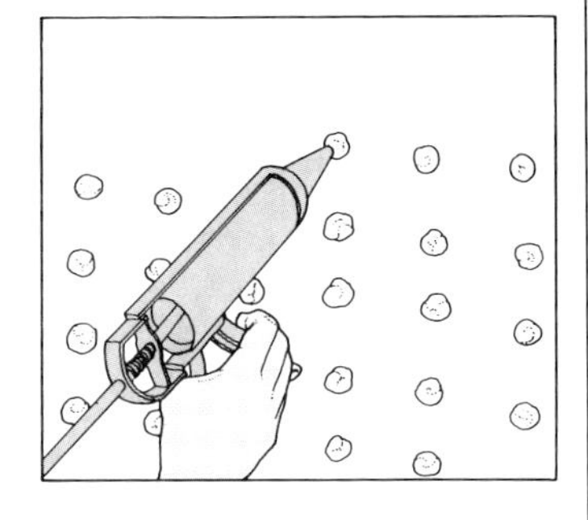

Having drawn a line approximately where the sides and top of the mirror will be, apply the glazing compound to the wall in lumps about the size of walnuts. They should be placed in rows 50mm to 75mm apart.

Then apply lines of silicone between the rows of compound.

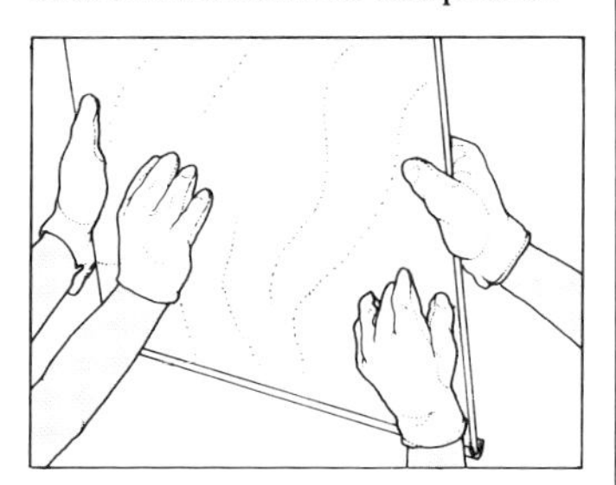

Now, with helpers, place the bottom of the mirror on its support and gently push it back against the compound, applying pressure evenly over the entire surface. Continue until it won't go any further.

The compound is non-setting and the silicone takes about 24 hours to set, so some re-positioning of the mirror is possible. This should be done immediately and movement kept to a minimum.

MIRROR TILES & MOSAICS

The effects and benefits of using panels of mirror can also be produced with mirror tiles. They can be used together to create as large a mirror as you like. Mirror tiles do not give the same quality of image as large mirrors; their reflection is broken into squares. This creates an interesting impressionist effect.

In situations where a bright, clear reflection isn't needed, bronze or smoked grey mirror tiles will produce a subdued and sophisticated atmosphere.

Mirror tiles are less expensive to use than large panels. Square mirror tiles are usually available in 108mm, 150mm, 225mm, and 300mm sizes. Rectangular sizes are usually 300mm × 150mm, although it is possible to order tiles up to 600mm. Shaped tiles are also available in circles, semi-circles and quadrants.

Mirror tiles are suitable for use in kitchens and any other room where extra light is needed. Properly installed they won't be affected by moisture and are as easy to wipe clean as ceramic tiles. The kitchen is one place in the home where as much light as possible is needed to prevent eyestrain and accidents when preparing food, and mirrors will increase the usable light.

An entire wall covered with mirror tiles can double the apparent size of the room and increases the level of natural light. Objects reflected create an attractive three-dimensional effect. Strategically placed plants help to conceal the presence of the mirrors and allow the occupants to enjoy the feeling of extra light and space without being continually aware of their own reflections.

Unusual effects can be produced by using mosaic mirror or mirrorflex. This is a fabric material holding a mosaic of small mirror squares. As well as providing a reflective surface, mirrorflex creates interesting decorative effects. It can be bought in sheets and cut to the required size and shape, and also can be used around corners and on pillars and columns.

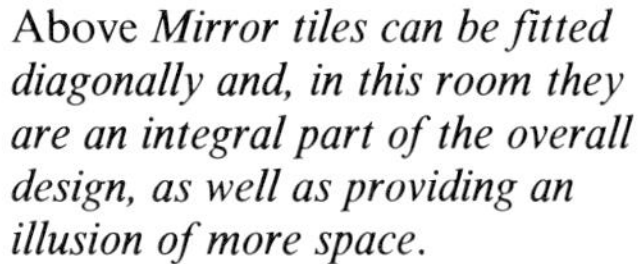

Above *Mirror tiles can be fitted diagonally and, in this room they are an integral part of the overall design, as well as providing an illusion of more space.*

Above right *Plants help to camouflage the use of mirror tiles in this small room. Notice how much more interesting the room appears, and how the tiles in no way detract from the homely cosy feel. Mosaic tiles have been used to form a border.*

Right *A plain door is brought alive with mirrorflex. As light hits these tiny mirrors, exciting light patterns are produced.*

Left *The reflections of the painted bricks and floor tiles give an illusion of space to an otherwise small room. Placed opposite the garden and patio doors, the mirror tiles provide interesting, attractive reflections.*

Fixing Mirror Tiles

Mirror tiles can be fixed to any clean, dry, flat non-porous surface. Porous surfaces, such as plaster, wood or emulsion paint, must be sealed with sealer or gloss paint — not vinyl. Newly plastered walls and recently applied ceramic tiles must be allowed to dry out thoroughly.

If you are covering a papered wall, remove the wallpaper from the areas where the tile's adhesive tabs will be applied. Be sure to seal the exposed patches.

Never apply tiles to cold surfaces. Unheated rooms and external walls should be heated first.

Clean the surface and ensure it is free from dust and check for flatness with a straight edge. If the wall is not flat, don't proceed. This would result in a clumsy, distorted reflection. If necessary, build a flat surface with timber battens and blockboard (see page 16).

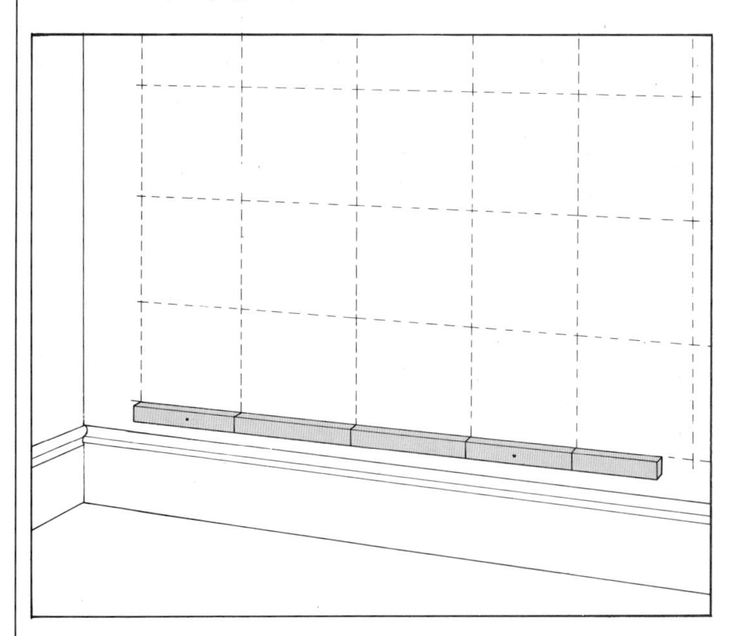

Fixing

As with all tiling, it's essential that all the mirror tiles are perfectly aligned on the wall. Make a tile gauge that corresponds to the size of your tiles and try it on the wall, marking the vertical and horizontal rows. By trial and error, find the best arrangement, hopefully requiring no cutting of the tiles. However, if part tiles at the edges are necessary, they should be equal at each side and top with bottom. Also the part tiles should be at least one half the tile width. If the edge tiles are less than this, reduce the number of tiles in the row by one.

A small gap should be left at the junction with adjacent walls and also at the ceiling.

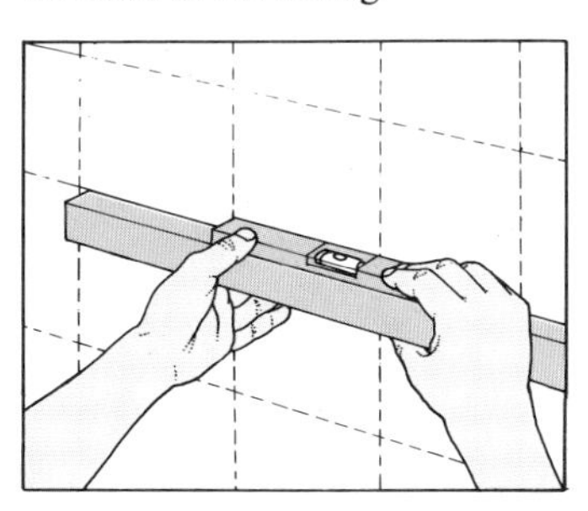

When you've settled on the best arrangement, fix a straight batten horizontally along one of the lines of tiles. Use a spirit level to make sure the batten is level and then check with your tile gauge that all your tiles will still fit.

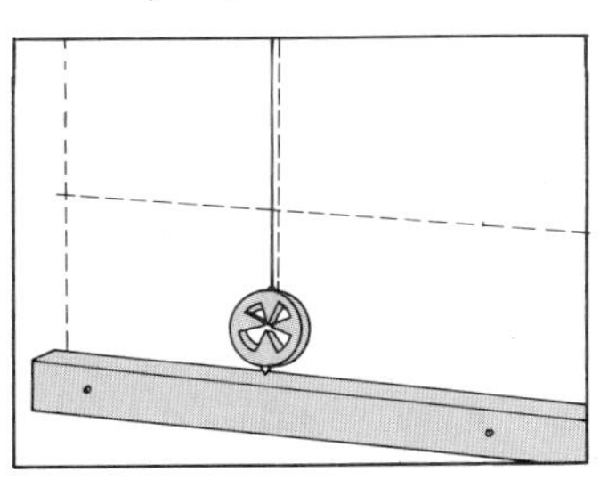

Next, using a plumb line, mark a vertical line corresponding to one of the vertical rows. Again, check that your arrangement is still o.k.

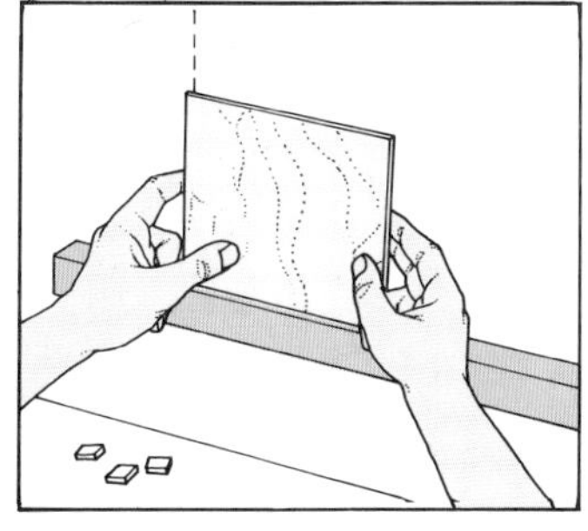

Peel the backing off the tabs on the tile and place it on the batten without touching the wall. Once you're sure the edge is in line with the vertical line, press the tile back against the wall. Be sure to get the position right first time, as any repositioning will spoil the adhesive power of the tabs. It's a good idea to have extra tabs handy in case of any slip.

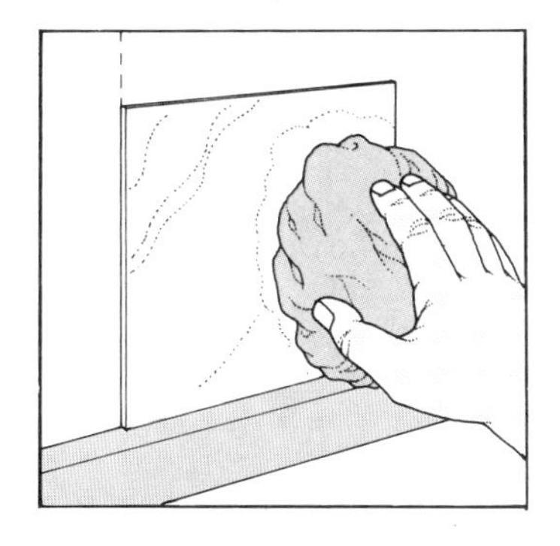

Use a soft cloth to press the tile firmly to the wall, pushing against each of the fixing tabs.

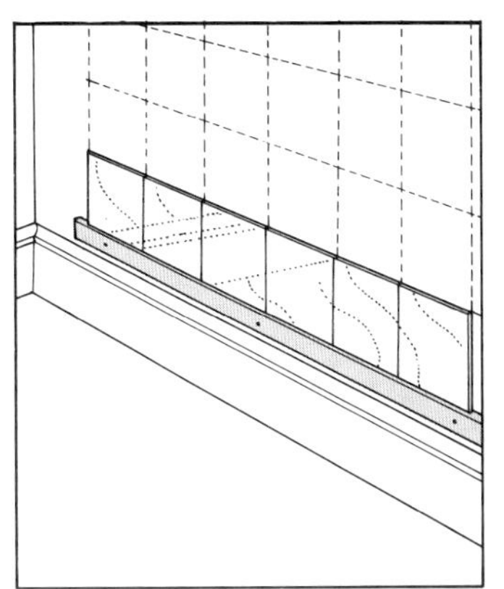

Continue applying tiles until the first horizontal row of whole tiles is completed. Follow the manufacturer's instructions as to whether the tiles should be butted together or a slight gap left between them.

Once the first row is complete, you can remove the batten, but be very careful not to disturb the tiles.

Fix the remaining tiles in horizontal rows.

Once the whole tiles are fixed, measure the gap at the edges remembering to leave a slight gap with adjacent walls and ceiling. Then take the edge tiles to the glass merchant for cutting or cut them yourself.

HOW MANY TILES?

Measure the height and width of the area to be tiled and use the chart to calculate how many tiles you will need. Buy a few extra tiles to allow for breakages if you are cutting.

Number of 15 cm × 15 cm tiles required

Room Height \ Room Width	30cm	60cm	90cm	1·2m	1·5m	1·8m	2·1m	2·4m	2·7m	3·0m	3·3m	3·6m
2·7m	36	72	108	144	180	216	252	288	324	360	396	432
2·4m	32	64	96	128	160	192	224	256	288	320	352	384
2·1m	28	56	84	112	140	168	196	224	252	280	308	336
1·8m	24	48	72	96	120	144	168	192	216	240	264	288
1·5m	20	40	60	80	100	120	140	160	180	200	220	240
1·2m	16	32	48	64	80	96	112	128	144	160	176	192
90cm	12	24	36	48	60	72	84	96	108	120	132	144
60cm	8	16	24	32	40	48	56	64	72	80	88	96
30cm	4	8	12	16	20	24	28	32	36	40	44	48

Room Width

MIRROR WARDROBE DOORS

Wardrobes with sliding mirror doors are mirrored walls in their most practical and useful form. They give you all the benefits of floor to ceiling mirror panels (doubling space and increasing light, etc.) and they conceal a large area of bedroom storage as well. So, although the cupboards reduce the size of the room, the mirrors make it seem larger than before.

Mirrors on wardrobe doors fall into three main categories. All can be used with existing cupboards.

There are systems of tracks and sliding mirror panels sold in standard sizes that you can combine to span any distance. In these the mirrors are the doors – there is nothing behind the glass apart from the protective safety film that all such mirrors must have. Don't confuse the DIY backing used to protect the back of the mirror with this safety film. It is commercially applied and should be used for all mirrors in situations where people could fall against them.

A second way to use mirror is to fix mirror panels, mirror tiles or mosaics to the fronts of existing doors. You could use either mirror clips or screw fixing for this. The mirrors could cover the doors completely, or you can leave a border of the doors showing. If the mirrors are to come right to the edge of the doors, remember to have holes drilled for the handles when ordering them.

Another idea is to use mirrors in a way that suggests the panelling in traditional doors. On flat doors you can fix the panels in the normal way and then frame them with panel mouldings. If your doors are of traditional frame and panel construction, remove the panel moulding and replace it after fixing the mirrors onto the panels.

Right *A typical example of the sliding mirror doors readily available in kit form.*

Below left *The black borders of the existing doors set off the drilled mirror panels that have been screwed to them.*

Below *These traditional-style fitted cupboards are brought to life by the addition of mirrors in the panels, adding the illusion of space and light to the room.*

MIRRORS IN FRAMES

Ceramics, bamboo, pine, polished oak, brass, gilt — the list of materials that can be used for framing mirrors is endless. The mirror glass itself is equally versatile. Edges can be bevelled and the surface can be brilliant cut, acid etched or sandblasted. The combination of an attractive frame and a skillfully finished mirror can transform the functional looking glass into a focal point for a room.

Whether you should replace an antique mirror glass really depends on its value. If the silver is badly stained or degraded and the piece isn't worth much, you can replace it with new mirror to good effect. However, if you want the effect of old glass, look around for a suitable piece larger than the frame you have, and ask the glass merchant to cut it to size for you.

Re-silvering is not really a do-it-yourself job. The materials and machinery needed to produce an effective, lasting backing of a high standard make this the province of professional restorers.

Above *A plain framed mirror (with the addition of glazing bars) placed in a narrow corridor adds a new dimension to the view from within the room.*

Above right *The gilt frame of this large mirror was badly damaged and has been painted white to match the decor. It now makes an eye-catching centre piece for this basement room and introduces much needed reflected light.*

Right *A framed mirror compliments this nostalgic bathroom suite. Prints and drawings on the opposite wall are reflected, filling the bare area above the wash basin with interest and colour. Unlike the artwork, the mirror's surface won't be affected when splashed with water.*

Framing a Mirror

There are two ways to frame a mirror. You can make or buy a frame, fit a mirror into it, and then hang it like a picture. Alternatively, you may want to surround a large mirror panel already fixed to a wall. The making of picture frames is covered in the companion book **Frame It**.

Second-hand Frames

The first step is to remove the brads holding the picture in place.

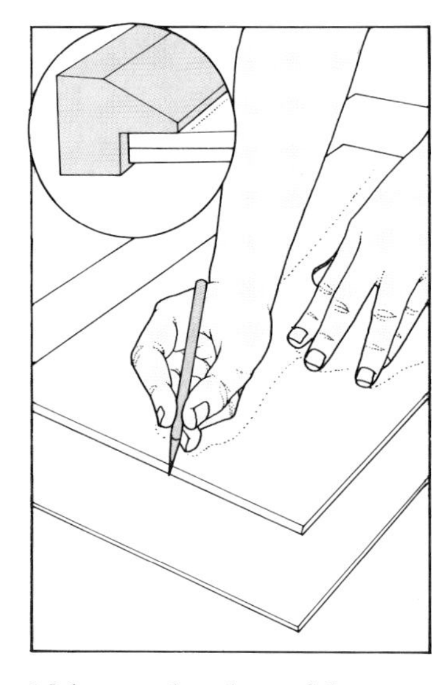

Make sure the rebate of the frame is deep enough to accept the thickness of mirror and 3mm hardboard. The glass merchant will advise the correct thickness of mirror depending on the overall size.

Measure the dimensions of the inside of the frame and order a mirror from your glass merchant of a size that will leave 2mm or so between mirror and frame at each edge. Also, you will need a piece of 3mm hardboard cut to the same size. If the local timber merchant doesn't offer this service, you can cut it yourself after tracing around the edge of the glass.

If the joints of the frame aren't in good condition and may not take the weight of the mirror, you can strengthen them with metal plates on the back of the frame.

Place the mirror in the frame and put the hardboard in on top. This will protect the back of the mirror. Then, use brads to hold the hardboard snugly in the frame.

Surrounds

To surround a fixed mirror, you can choose from the many wood mouldings and picture frame mouldings that are available. Some of these are pre-finished in lacquer, paint, gold leaf, etc. and they may be made of metal, wood or plastic.

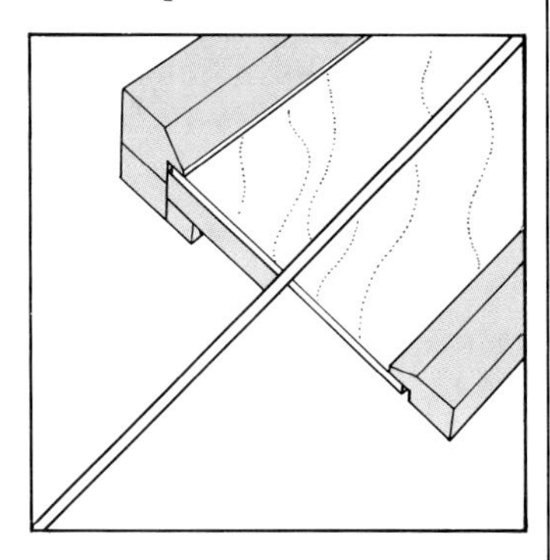

If your mirror is mounted on a blockboard base, fixing the frame around it will be easy. Simply fix strips of wood to cover the edge of the blockboard, and pin the frame on the front.

If, on the other hand, the mirror is fixed directly to the wall, your choice of moulding will be more critical as the rebate in the moulding will need to be shallow enough for the edge to overlap the glass.

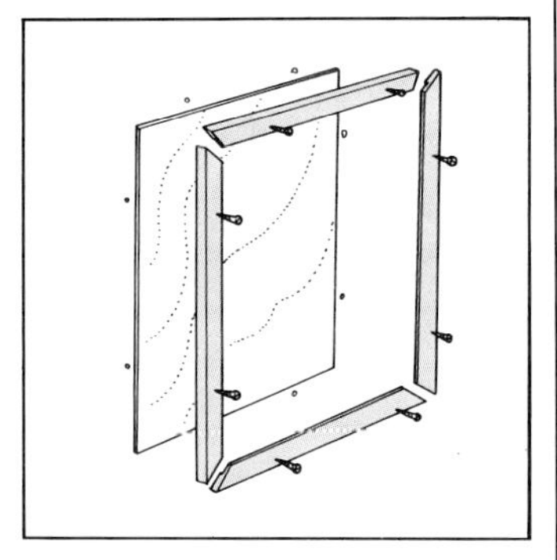

One method of fixing is to drill and screw the moulding to the wall, but be careful not to put undue stress on the edge of the mirror. The holes in the frame can then be filled and decorated with the rest of the frame.

UNUSUAL USES

The reflective surface and light-bouncing properties of mirrors enable exciting results to be achieved in home decor. Furniture, shelves, dado rails and panels can each provide a strong, secure base for installing a mirror. By using mirrors cunningly around the home, you can amuse and deceive visitors with exciting trompe l'oeil effects.

Mirrors can create an illusion of space in a compact garden; not only is the feeling of space produced, but the impact of the plants and shrubs is doubled.

Above *Mirror panels can be tucked away almost anywhere, brightening up the decor and, in this case, disguising a standard fire check door.*

Right *A mirror has been craftily fitted into this archway to produce the illusion of looking through to another garden. The entrance is blocked by shrubs. This will prevent visitors, who have fallen for the illusion, walking into the mirror.*

Above left *A simple panel of mirror, placed around a breakfast bar or kitchen counter, breaks up a monotonous partition for an attractive three-dimensional effect.*

Above *Mirror, fitted into an alcove behind glass shelves, makes attractive, bright and decorative storage.*

Left *An old overmantel, painted white, makes an appealing and useful bathroom mirror.*

Top Ten Tips & Safety Tips

1 When fixing mirror tiles to embossed ceramic tiles,use 50% more fixing pads than are already fixed on the back of the mirror. This is to compensate for the loss of adhesive contact caused by the embossing and grouting. Extra pads can be obtained from the mirror tile supplier.

2 Mirrors used outdoors should be sealed to protect the silver from moisture. Use mirror with safety backing and apply silicone to the edges. Masking tape will keep the silicone off the front of the mirror. Peel it off before the silicone sets.

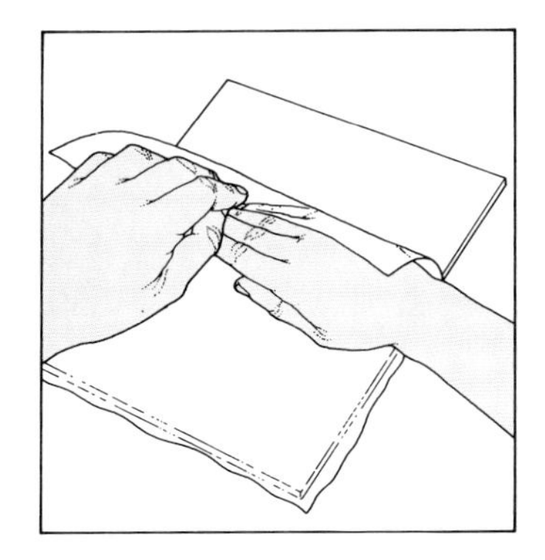

3 A simple way to make small pieces of mirror safer is to stick self-adhesive plastic on the back. If the mirror breaks the pieces will stick to the backing.

4 Modern mirror quality is high, but before buying check for these possible blemishes — bubbles, blisters, scratches, blobs and embedded particles.

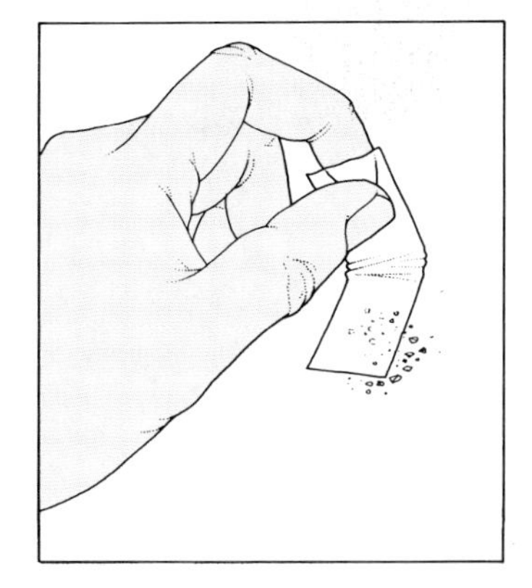

5 Use masking or sticky tape to pick up those tiny slivers of mirror (there is now a purpose-made roller in the haberdashery department of large stores).

6 Always clean up broken mirror or glass immediately. Apart from large cuts, the tiny pieces can be dangerous to pets and children.

Mirror Maintenance

Mirror, although expensive to buy, should last a lifetime if properly installed and maintained. Mirrors should be cleaned regularly to keep them in good condition. Here are a few points to remember.

The most vulnerable point of a mirror is the edge of the silver backing. This is because some liquids can creep between the silver and glass; also, the backing is fragile and the edge is easily damaged if care is not taken.

Don't use dirty or gritty rags, knives, scrapers, emery cloths or any abrasive material for cleaning.

Abrasives, alkali and *acids* — the three 'A's' — are harmful to mirrors and must be avoided. Never use heavy-duty commercial cleaning solutions that usually contain one or more of these substances.

Discuss with your mirror supplier the most suitable cleaner. Here are three solutions that can be used:

1. Weak (5–10%) *alcohol in water* solution
2. Weak (5%) household *vinegar in water* solution
3. Weak household *ammonia in water* solution

It is important that these are used sparingly and not allowed to come in contact with the edges or back of the mirror. Only the face should be cleaned with solution — the edges are best cleaned with a dry cloth. One way to avoid wetting the edges is to apply the cleaning solution to the cloth, rather than the glass.

The same applies when using any of the various proprietary glass and mirror cleaners. When using a spray or aerosol, be careful not to spray directly on the edge.

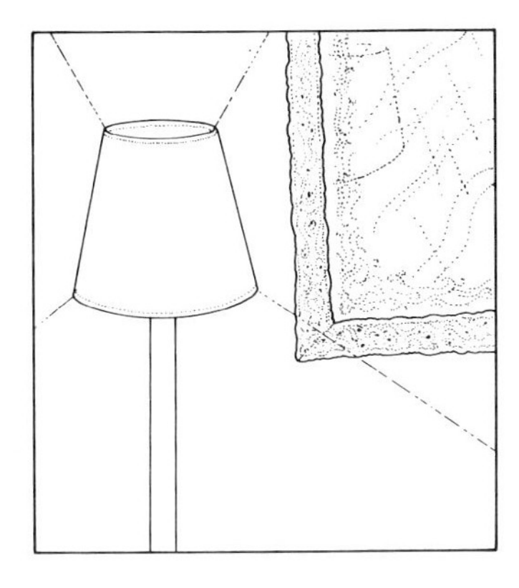

7 Don't use direct light on an old mirror — it accentuates the flaws. Rely on reflected light, which is softer.

8 Check that the large mirror you have ordered will fit through the front door and up the stairs!

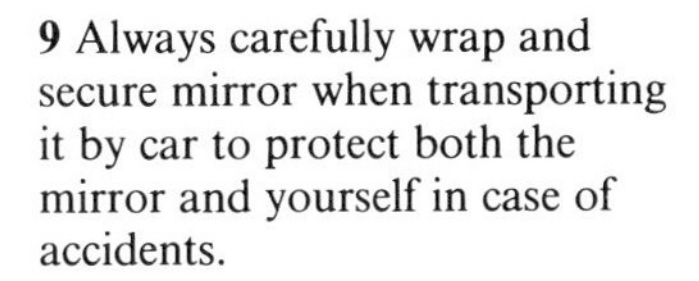

9 Always carefully wrap and secure mirror when transporting it by car to protect both the mirror and yourself in case of accidents.

10 If you have young children, think before placing a mirror over a working fireplace. Children stretching up to see their reflections could set fire to their clothes.

Add a touch of glamour to your cloakroom with theatrical-type lighting. (Be sure to follow wiring instructions carefully.)

Reflected light and mirror help cut-out the claustrophobic feeling sometimes felt in small cloakrooms.

Useful Addresses

Britain
Glass and Glazing Federation
44–48 Borough High St.
London SE1 1XP
Telephone 01-403 7177

Australia
Glass Merchants Association
Site 12A
100 Drummond St.
Victoria 3053

Canada
Glazing Contractors' Ass.
4299 Canada Way
Burnaby B.C. ZESGIH3

New Zealand
Auckland Glass Merchants Association
196–198 Federal St.
Auckland 1
Double Glazing Manufacturers' Association
247 Remuera Road
Remuera, Auckland

USA
National Glass Association
8200 Greensboro Drive
Route 302, McLean,
VA 22102

Author
Ian Muir
Series Consultant Editor
Bob Tattersall
Editor
Dek Messecar
Design
Bob Lamb
Picture Research
Liz Whiting
Illustrations
Rob Shone

Ian Muir is public relations officer of the Glass and Glazing Federation. In researching and writing this book he was able to draw upon the help and expertise of members of the GGF Mirror Advisory Group.

Bob Tattersall has been a DIY journalist for over 25 years and editor of *Homemaker* for 16 years. He now works as a freelance journalist and broadcaster. Regular contact with the main DIY manufacturers keeps him up-to-date on all new products and developments. He has written many books on various aspects of DIY and, while he is considered 'an expert', he prefers to think of himself as a do-it-yourselfer who happens to be a journalist.

Picture Credits
Photographs from Elizabeth Whiting Photo Library by Jon Bouchier, Steve Colby, Michael Crockett, Richard Davies, Michael Dunne, Clive Helm, Home Improvements, Michael Nicholson, Spike Powell, Tim Street-Porter, Jerry Tubby, Glass & Glazing Federation.

Cover Photography by Clive Helm

The Do It Series was conceived, edited and designed by Elizabeth Whiting & Associates and Robson Lamb & Company for William Collins Sons & Co Ltd

First published 1987
9 8 7 6 5 4 3 2 1

Published by William Collins Sons & Co Ltd
London · Glasgow · Sydney · Auckland
Toronto · Johannesburg

ISBN 0 00 412191 0

Photoset by V & M Graphics Ltd, Aylesbury, Bucks
Colour separations and printed in Singapore
by Tien Wah Press (Pte) Ltd